THE 100+ SERIES™

ALGEBRA II

Essential Practice for Advanced Math Topics

Carson Dellosa Education
Greensboro, North Carolina

Visit carsondellosa.com for correlations to Common Core, state, national, and Canadian provincial standards.

Carson Dellosa Education
PO Box 35665
Greensboro, NC 27425 USA
carsondellosa.com

ISBN 978-1-4838-0078-3

09-011217784

Table of Contents

Introduction

What are the Common Core State Standards for Middle School Mathematics?

In grades 6–8, the standards are a shared set of expectations for the development of mathematical understanding in the areas of ratios and proportional relationships, the number system, expressions and equations, functions, geometry, and statistics and probability. These rigorous standards encourage students to justify their thinking. They reflect the knowledge that is necessary for success in college and beyond.

Students who master the Common Core standards in mathematics as they advance in school will exhibit the following capabilities:

1. Make sense of problems and persevere in solving them.

Proficient students can explain the meaning of a problem and try different strategies to find a solution. Students check their answers and ask, "Does this make sense?"

2. Reason abstractly and quantitatively.

Proficient students are able to move back and forth smoothly between working with abstract symbols and thinking about real-world quantities that symbols represent.

3. Construct viable arguments and critique the reasoning of others.

Proficient students analyze problems by breaking them into stages and deciding whether each step is logical. They justify solutions using examples and solid arguments.

4. Model with mathematics.

Proficient students use diagrams, graphs, and formulas to model complex, real-world problems. They consider whether their results make sense and adjust their models as needed.

5. Use appropriate tools strategically.

Proficient students use tools such as models, protractors, and calculators appropriately. They use technological resources such as Web sites, software, and graphing calculators to explore and deepen their understanding of concepts.

6. Attend to precision.

Proficient students demonstrate clear and logical thinking. They choose appropriate units of measurement, use symbols correctly, and label graphs carefully. They calculate with accuracy and efficiency.

7. Look for and make use of structure.

Proficient students look closely to find patterns and structures. They can also step back to get the big picture. They think about complicated problems as single objects or break them into parts.

8. Look for and express regularity in repeated reasoning.

Proficient students notice when calculations are repeated and look for alternate methods and shortcuts. They maintain oversight of the process while attending to the details. They continually evaluate their results.

How to Use This Book

In this book, you will find a collection of 100+ reproducible practice pages to help students review, reinforce, and enhance Common Core mathematics skills. Use the chart provided on the next page to identify practice activities that meet the standards for learners at different levels of proficiency in your classroom.

Common Core State Standards* Alignment: Algebra II

Domain: Seeing Structure in Expressions	
Standard	**Aligned Practice Pages**
HSA-SSE.A.1a	11–13
HSA-SSE.A.1b	11, 20
HSA-SSE.A.2	22–28
HSA-SSE.A.3b	46

Domain: Arithmetic With Polynomials and Rational Expressions	
Standard	**Aligned Practice Pages**
HSA-APR.A.1	7–10, 14–21, 24–28
HSA-APR.B.2	22–28, 35, 36
HSA-ARP.B.3	37, 38
HSA-ARP.D.6	29–36
HSA-APR.D.7	29–36

Domain: Reasoning With Equations and Inequalities	
Standard	**Aligned Practice Pages**
HSA-REI.A.2	39, 40
HSA-REI.B.3	41, 42
HSA-REI.B.4b	41, 42
HSA-REI.D.10	43–45
HSA-REI.D.11	43–45

Domain: Linear, Quadratic, and Exponential Models	
Standard	**Aligned Practice**
HSF-LE.A.2	43–45, 53, 84, 88, 91
HSF-LE.A.4	51–53, 55, 56

Domain: Interpreting Functions	
Standard	**Aligned Practice Pages**
HSF-IF.A.1	57, 58
HSF-IF.B.4	48
HSF-IF.B.5	57, 58
HSF-IF.C.7a	46–49, 84, 87, 89, 90, 92, 93, 95
HSF-IF.C.8a	85, 94

Domain: Building Functions	
Standard	**Aligned Practice Pages**
HSF-BF.A.2	85, 86, 88, 90, 91, 93, 94
HSF-BF.B.5	54–56

Domain: Trigonometric Functions	
Standard	**Aligned Practice Pages**
HSF-TF.A.1	65–67
HSF-TF.A.2	65–67
HSF-TF.A.3	65–70
HSF-TF.B.6	72–80
HSF-TF.B.7	72–80
HSF-TF.C.8	68–70
HSF-TF.C.9	71

Domain: Similarity, Right Triangles, and Trigonometry	
Standard	**Aligned Practice Pages**
HSG-SRT.C.6	60–64
HSG-SRT.C.7	61–64
HSG-SRT.D.10	81–83
HSG-SRT.D.11	81–83

Domain: Vector and Matrix Quantities	
Standard	**Aligned Practice Pages**
HSN-VM.A.1	96
HSN-VM.A.2	99, 100
HSN-VM.A.3	101–103
HSN-VM.B.4a	97, 98
HSN-VM.B.4b	97
HSN-VM.B.5a	98

Name______________________________

HSA-APR.A.1

Adding and Subtracting Polynomials

$(x^3 + 2x^2 - 8x) - (^{-}2x^2 + 7x - 5) = x^3 + 2x^2 - 8x + 2x^2 - 7x + 5 = x^3 + 4x^2 - 15x + 5$

1. $(4x + 2) + (x - 1) =$
2. $(5a - 2b + 4) + (2a + b + 2) =$
3. $(3a + 2b) - (a - b) =$
4. $(x^2 + y^2 - ab) - (x^2 - y^2 + ab) =$
5. $(4a^2 - 5ab - 6b^2) + (10ab - 6a^2 - 8b^2) =$
6. $(4x^2 - 2x - 3) - (^{-}5x - 4) =$
7. $(4a^2 - 4ab - b^2) + (a^2 - b^2) + (2ab + a^2 + b^2) =$
8. $(^{-}4x^3 - 6x^2 + 3x - 1) - (8x^3 + 4x^2 - 2x + 3) =$
9. $(a + 2b) + (3b - 4c) + (5a - 7c) + 3b =$
10. $(x^2 - 2xy + y^2) - (x^2 - 2xy + y^2) =$
11. $(x + 3y) + (^{-}3x - y) - (x - y) =$
12. $(2x^2 + 3y^2 - z^2) - (x^2 - y^2 - z^2) + (4x^2 - 3y^2) =$
13. $(2x + 3) + (^{-}2x^2 + x - 5) =$
14. $(2y + 3x - 4) + (9 - 8y - 5x) + (3x + 4y - 2) =$
15. $(^{-}2y^2 + 8) - (3y^2 - 4y - 6) =$
16. $(7y + 4x + 9) - (6x - 8y + 11) =$

Find the perimeter.

17.

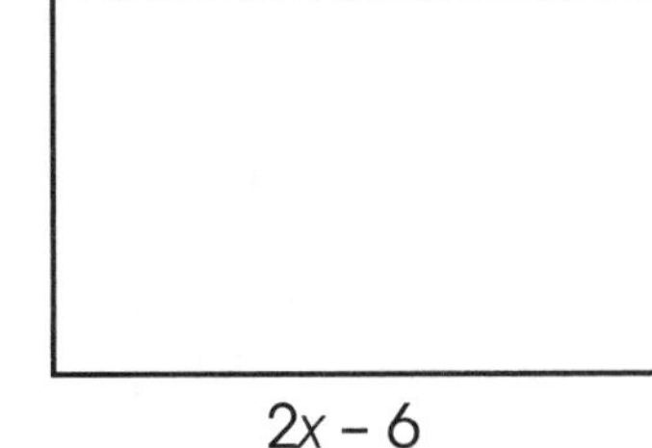

18.

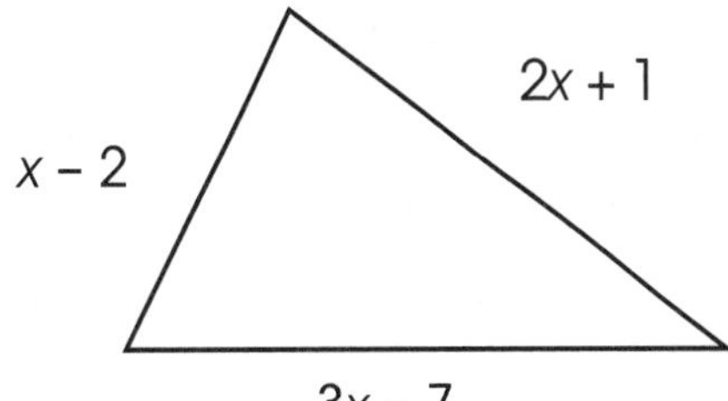

Name________________________________

HSA-APR.A.1

Solving Geometry Problems with Polynomials

Find the perimeter of each polygon.

1.

2.

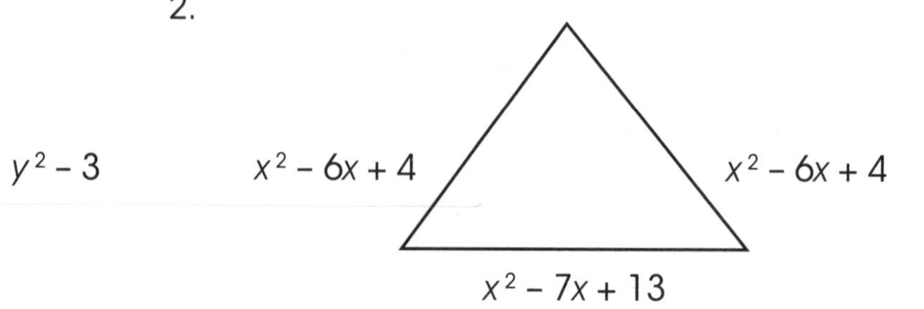

3.

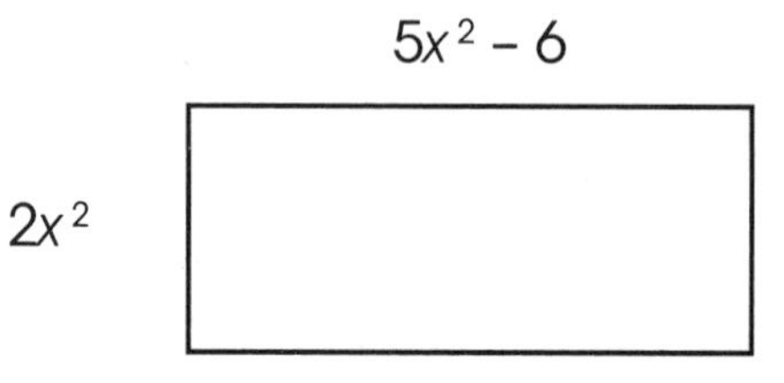

4.

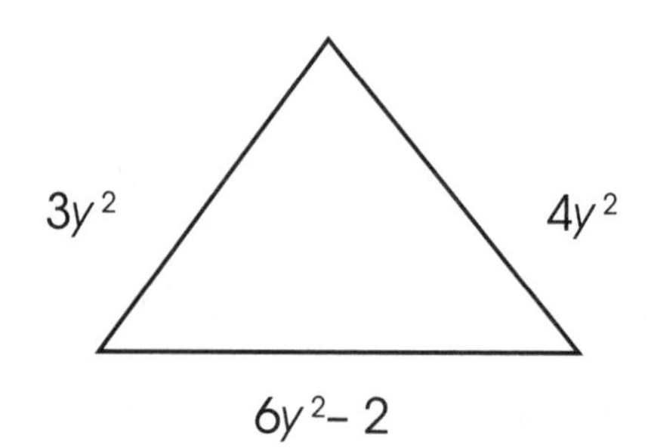

Find the area of each polygon.

5. $A = \frac{1}{2}\, b \cdot h$

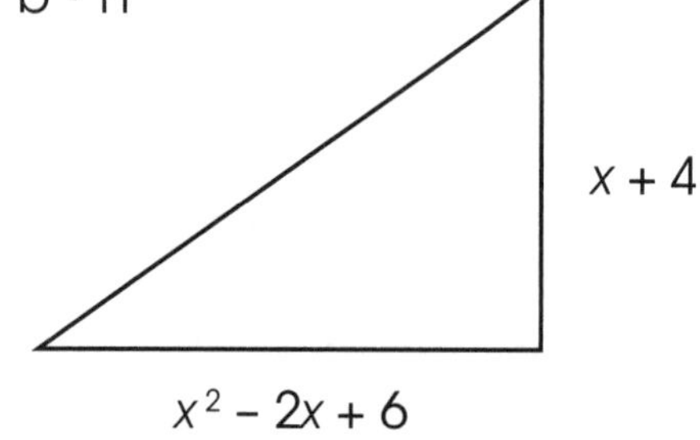

6. $A = s^2$

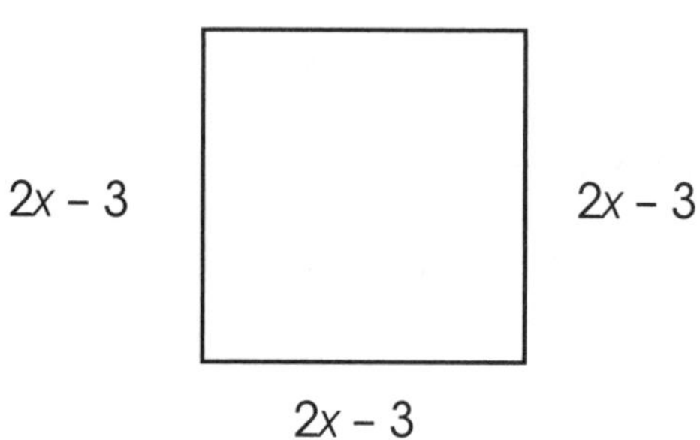

7. $A = l \cdot w$

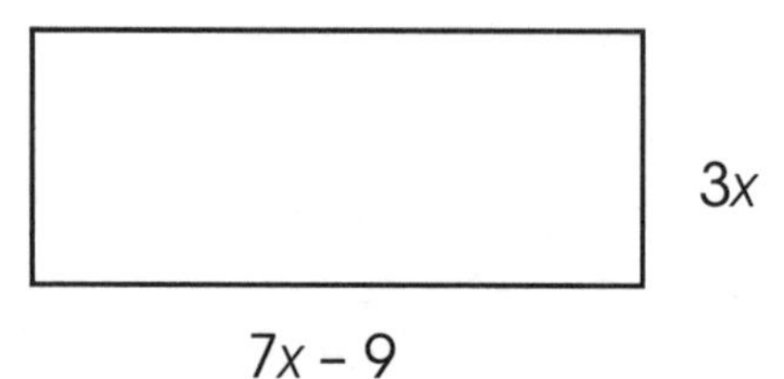

8. $A = s^2$

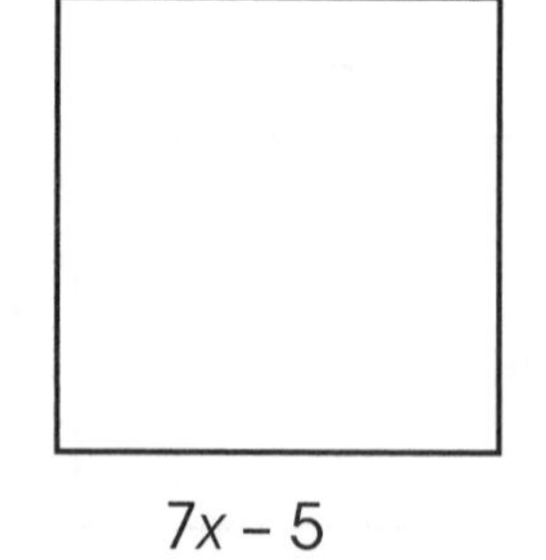

Name________________________________ HSA-APR.A.1

Solving Problems with Polynomials

Sally has 21 coins in nickels and dimes. Their total value is \$1.75. How many of each coin does she have?

Let x = number of nickels.

$21 - x$ = number of dimes

$$5x + 10\,(21 - x) = 175$$

value of nickels — value of dimes — total value in cents

$$5x + 210 - 10x = 175$$

$$210 - 5x = 175$$

$$^-5x = {^-35}$$

$$x = 7$$

There are 7 nickels and 14 dimes.

1. Tim bought some 25¢ and some 29¢ stamps. He paid \$7.60 for 28 stamps. How many of each type of stamp did he buy?

2. Tickets for the school concert were \$3 and \$2. If 245 tickets were sold for a total of \$630, how much of each kind were sold?

3. David has 11 coins, some quarters and some dimes. If the coins have a value of \$1.55, how many of each kind are there?

4. Subtract $6x^2 - 3xy + y^2$ from $8x^2 + 5xy - y^2$.

5. From $6ab - 2ac + 5bc$, take $10ab - 2bc + 3ac$.

6. Subtract $x^2 - y^2 - z^2$ from the sum of $3x^2 + 2y^2 + z^2$ and $4x^2 + 3y^2 - 5z^2$.

7. Find the perimeter of a rectangle if its length is $(4c + 7)$ units and its width is $(c - 3)$ units.

8. Which has the greater area, a square with sides each $(x + 2)$ units long or a rectangle with length $(x + 4)$ units and width x units?

Name________________________________

Polynomials Equations

$$2(2x - 1) + 5x = 3(x - 2) + 10$$
$$4x - 2 + 5x = 3x - 6 + 10$$
$$9x - 2 = 3x + 4$$
$$6x - 2 = 4$$
$$6x = 6$$
$$x = 1$$

1. $5(x + 2) - 4(x - 1) = 24$
2. $2(y - 5) - (y + 6) = {}^{-}4$
3. $2(3n - 1) - (n + 6) = 7$
4. $2(2b + 7) + 2(1 + 2b) = 20$
5. $(4x - 1) - (2x + 2) = x + 5$
6. $2(y + 1) + 3(y - 1) = 9$
7. $(x + 1)(x + 5) = (x + 2)(x + 3)$
8. ${}^{-}4r + 3(1 - 2r) = 3(5 - 2r)$
9. $(y + 12)(y - 3) = y(y + 5) + 24$
10. $(2x - 3)(2x - 1) = (x - 2)(4x + 3)$
11. $(x - x^2) - (2x^2 + x - 1) = 5 + 2x - 3x^2$
12. $t(t - 2) + 2(2t - 1) = t^2 + 4$
13. $(w + 4)(w + 14) - w(w + 10) = 216$
14. $(2x + 1)^2 - (2x - 1)^2 = (x + 6)^2 - x^2$

Name________________________________ HSA-SSE.A.1a, HSA-SSE.A.1b

Monomial Quiz

Polly N. Omial did not understand the rules of exponents when she completed the Monomial Quiz. Find and correct the 9 errors Polly made.

Monomial Quiz Name: *Polly*

1. $(4c)^2 = 8c^2$
2. $4(c)^2 = 4c^2$
3. $(2a^2b)(4ab^2) = 8a^3b^3$
4. $(4pq)(-p^2q^3) = {}^-4p^3q^4$
5. $2x(-xy)(-y^2) = 2x^2y^3$
6. $(4bc)^2 = 16bc^2$
7. $-abc^2(cd)^3 = -abc^5d^3$
8. ${}^-2(3x)^2(xy)(2x)^2 = {}^-12x^5y$
9. $a(2a^2)^3 = 6a^6$
10. $3s({}^-2st)^2 = {}^-12s^2t^2$
11. $(-xy^2)^3(2x^2y)^2 = {}^-4x^7y^8$
12. $x^2({}^-2xz)(4z^5) = {}^-3^2x^3z^6$
13. $(3pq^2r^3)\dfrac{(q^2r)}{3} = pq^4r^4$
14. $(-x)({}^-2xy)({}^-3xyz) = {}^-6x^3y^2z$
15. $2x^2(xy^2)^2(xz^2)^2 = 2x^4y^4z^4$
16. $(2u)^2(u^2v)^3(w) = 2u^4v^3w$
17. $(-x)(-x^5) = x^6$
18. $({}^-5x^2)(7xy^3) = {}^-35x^3y^3$
19. $(-x^3y)(6xy)^2 = {}^-6x^4y^3$
20. $({}^-4rt^2)(2rt)({}^-t^2) = 8r^2t^5$

Explain to Polly how to calculate the power of a power, for example, $({}^-5x^2)^3$.

____

____

Name________________________________

HSA-SSE.A.1a

Classifying Polynomials

General expression of a polynomial

$a_n x^n + a_{n-1} x^{n-1} + a_{n-2} x^{n-2} + \ldots a_1 x^1 + a_0 x^0$

Note: $x^0 = 1$

$a_0 \cdot 1 = a_0$

<u>Types of Polynomial</u>

monomial = 1 term

binomial = 2 terms

trinomial = 3 terms

1st degree = linear polynomial

2nd degree = quadratic polynomial

3rd degree = cubic polynomial

(Degree is the highest power on the variable.)

Classify each polynomial.

Example 1: $4x + 7$

linear binomial

Example 2: $7x^4 + 12x^2 + 79$

4th degree trinomial

1. $x^3 + 1$
2. $10x^5 + x$
3. x^{10}
4. $15x^6 + x^4 + 1$
5. $10 + x^6$
6. $10x^{15}$
7. $x^2 + 10x^1 + x^3$
8. $x^2 + 70x^4$
9. $1500x$
10. $17x^7 + 1$
11. $5x^6 + 4$
12. $21x^3 + 4x + 10$
13. $6 + 11x + x^4$
14. $6x + 4$

Name________________________________

HSA-SSE.A.1a

Ordering Polynomials

Write each polynomial in descending order including missing degrees of x.

Example 1: $5x^3 + 0x^2 + x + 4$
$5x^3 + x + 4$

Example 2: $9x + 6x^5 + 2x$
$6x^5 + 0x^4 + 0x^3 + 0x^2 + 11x + 0$

1. $7x^4 + x^3 + x$

2. $23 + 6x^3 + x^2$

3. $19x^6 + 1$

4. $4x + 14x^2$

5. x^3

6. $10x^1 + 3x^2$

7. $1 + 2x + x^2$

8. $8x^3 + 4x$

9. $1 + x$

10. $x + 3x^3 + 2x^2$

11. $18 + 20x + x^4$

12. $25x^2 + 5x$

13. $18x^3 + x$

Name________________________________

HSA-APR.A.1

Evaluating Polynomials

Given: $P(x) = 4x^2 + 1$
$Q(x) = {}^-2x^3 + x^2 - 6$
$R(x) = {}^-9x^2 - x + 16$

Evaluate each polynomial expression.

Example: $P(2)$
$P(2) = 4(2)^2 + 1$
$= 4(4) + 1 = 17$

1. $Q(5)$

2. $R(6)$

3. $Q({}^-2)$

4. $P(4)$

5. $R({}^-3)$

6. $R({}^-1)$

7. $P(10)$

8. $Q({}^-8)$

9. $P({}^-14)$

10. $Q(2)$

Cross out the correct answers below. Use the remaining letters to complete a statement, then rewrite the statement as a common adage.

101 SIM	-231 THE	-118 ILA	5 RAV	65 LAS	17 IAR	401 TMA	41 YSP	-18 NIS	37 ECI
-314 STA	-14 ESC	-62 NDI	341 ONG	-894 REG	1082 NG	-17 ATE	14 IST	8 HEO	785 NE.

_ _ _ _ _ _ _   _ _ _ _ _ _   _ _ _ _ _ _ _

_ _ _ _ _ _ _ _ _ _.

Common adage: ______________________________.

Name________________________________

HSA-APR.A.1

Multiplying a Polynomial by a Monomial

$$^{-}2a^2(9 - a - 4a^2) = {}^{-}2a^2 \cdot 9 - ({}^{-}2a^2 \cdot a) - ({}^{-}2a^2 \cdot 4a^2) = {}^{-}18a^2 + 2a^3 + 8a^4$$

$$(x + 2)(2x^2) = 2x^2 \cdot x + 2x^2 \cdot 2 = 2x^3 + 4x^2$$

1. $2(x^2 - xy + 3y^2) =$
2. $^{-}2n(4 + 5n^3) =$
3. $c^2d(c^2d^3 + 2cd^2 + d) =$
4. $2xy^2(2 - x - x^2y) =$
5. $(a^2 - 3ab - 2b^2)({}^{-}2ab) =$
6. $3n(8n^2 - 2n) =$
7. $(w^2z - 2wz + z)(-z^2) =$
8. $^{-}3ab^2(a^3b^2 - 2a^2b) =$
9. $4x^2y(9x^2 - 6xy^2 - 7) =$
10. $^{-}6k^2m^2(2k - 3m + 4km - k^2m^2) =$
11. $-n^2(n + 4n^2) =$
12. $(4x^2 - 7x)(-x) =$
13. $2x^2(x^3 - 2x^2 + 8x - 5) =$
14. $({}^{-}6x^3)(3x^2 - 1) =$
15. $(6x - 5x^2 + 8)({}^{-}3x) =$
16. $^{-}5x^2(2x^3 + 3x^2 - 7x + 9) =$

Find the area.

17. A = l • w

$x + 4$

$2x$

18. A triangle has a base length (b) of $2x + 4$ and a height (h) of $3y$. (Area = $\frac{1}{2}$ bh)

Name________________________

HSA-APR.A.1

Multiplying Polynomials

$$(s-2)(s^2-s+3) = s(s^2-s+3) - 2(s^2-s+3)$$
$$= s \cdot s^2 - s \cdot s + s \cdot 3 - 2 \cdot s^2 - 2(-s) - 2 \cdot 3$$
$$= s^3 - s^2 + 3s - 2s^2 + 2s - 6$$
$$= s^3 - 3s^2 + 5s - 6$$

1. $(z-3)(z+3) =$
2. $(3t-2)(t-3) =$
3. $(a+5)(a+5) =$
4. $(a+b)(2x+y) =$
5. $(\frac{1}{2}x - y)(2x+y) =$
6. $(4x-5)(4x+5) =$
7. $(1.6n-9)(0.2n-5) =$
8. $(2c+d)(c^2+2c+2d) =$
9. $(3a^2-2b^2)(3a^2+2b^2) =$
10. $(h+k)(h^2-2hk+3k^2) =$
11. $(2x-1)(x^2+x+3) =$
12. $(x^3+3x^2+2x-1)(x-1) =$
13. $(n-m)(n^2+m^2) =$
14. $(y+1)(y^2-2y+2) =$
15. $(\frac{1}{3}x-2)(\frac{1}{2}x+6) =$
16. $(3x^2-4x-7)(x+5) =$
17. $(x^2-3)(2x^2+3x+5) =$
18. $(4x^2-6x+4)(3x+2) =$

Name________________________________ HSA-APR.A.1

Multiplying Binomials Using FOIL

1. First, 2. Outer, 3. Inner, 4. Last

$(x + 5)(x - 3) = x \cdot x + x(^-3) + 5 \cdot x + 5(^-3) = x^2 - 3x + 5x - 15 = x^2 + 2x - 15$

1. $(x + 2)(x + 3) =$
2. $(y + 7)(y + 4) =$
3. $(x - 8)(x + 4) =$
4. $(x - 8)(x - 4) =$
5. $(y - 4)(y + 5) =$
6. $(x - 9)(x - 2) =$
7. $(2x + 4)(x + 3) =$
8. $(3x + 2)(2x + 5) =$
9. $(4x - 9)(3x + 1) =$
10. $(2x + 5)(4x - 3) =$
11. $(n - 7)(3n - 2) =$
12. $(5x + 2)(3x - 7) =$
13. $(^-4x + 5)(^-2x - 3) =$
14. $(-x - 4)(4 + 3x) =$
15. $(x + 2y)(2x + 3y) =$
16. $(6x - y)(3x - 2y) =$
17. $(4x + y)(3x - 4y) =$
18. $(5a + 3b)(4a - b) =$

Name________________________________

HSA-APR.A.1

Multiplication Match-Up

$(x + 2)(x - 4) = x^2 - 4x + 2x - 8 = x^2 - 2x - 8$

F O I L

Column A

1. $(x - 4)(x + 3)$
2. $(x + 2)(x - 6)$
3. $(3x + 1)(2x - 3)$
4. $(x + 3)(6x - 1)$
5. $(5x + 7)(5x + 7)$
6. $(4x - 1)(4x + 1)$
7. $(0.2x - 3)(0.3x - 2)$
8. $(3x + 1)(0.2x + 5)$
9. $(4x + 5)(6x + 7)$
10. $(8x - 5)(3x + 2)$

Column B

I. $25x^2 + 70x + 49$

T. $6x^2 - 7x - 3$

O. $x^2 - x - 12$

A. $24x^2 + x - 10$

L. $0.6x^2 + 15.2x + 5$

R. $x^2 - 4x - 12$

M. $6x^2 + 17x - 3$

S. $24x^2 + 58x + 35$

I. $0.06x^2 - 1.3x + 6$

N. $16x^2 - 1$

All but one of the answers are __ __ __ __ __ __ __ __ __ __.

3 2 5 6 1 4 7 10 8 9

What is special about the factored form of answer *N*—$16x^2 - 1$? ______________________

____

Name________________________________

HSA-APR.A.1

Special Products

$$(2x + 5)(2x - 5) = 4x^2 - 10x + 10x - 25 = 4x^2 - 25$$
$$(a + b)(a - b) = a^2 - b^2$$

1. $(x + 3)(x - 3) =$

2. $(y - 10)(y + 10) =$

3. $(a + 4)(a - 4) =$

4. $(x + 7)(x - 7) =$

5. $(2x + 1)(2x - 1) =$

6. $(5y - 6)(5y + 6) =$

7. $(4x + 3)(4x - 3) =$

8. $(3n + 7)(3n - 7) =$

9. $(3c + 4)(3c - 4) =$

10. $(2x + 9)(2x - 9) =$

11. $(7x - 5)(7x + 5) =$

12. $(x + y)(x - y) =$

13. $(5x - y)(5x + y) =$

14. $(2x + 11y)(2x - 11y) =$

15. $(3x - 7y)(3x + 7y) =$

16. $(2x + y^2)(2x - y^2) =$

17. $(3x^2 - 1)(3x^2 + 1) =$

18. $(x^2 + y^2)(x^2 - y^2) =$

Thinking About It

19. The product of the sum and difference of two terms is equal to the ______________ of the squares of the terms.

20. Why is the product of a sum and difference of two terms a binomial and not a trinomial?

____

Name________________________________

HSA-SSE.A.1b, HSA-APR.A.1

Squaring Binomials

$(a + b)^2 = (a + b)\ (a + b) = a^2 + ab + ab + b^2 = a^2 + 2ab + b^2$

1. $(x - 8)^2 =$

2. $(a + 5)^2 =$

3. $(x - 3)^2 =$

4. $(3n + 1)^2 =$

5. $(y - 10)^2 =$

6. $(3x + 2)^2 =$

7. $(4x - 3)^2 =$

8. $(2a + 5)^2 =$

9. $(6x + 1)^2 =$

10. $(5b + 2)^2 =$

11. $(4x - y)^2 =$

12. $(6x - 5y)^2 =$

13. $(3y - 5z)^2 =$

14. $(7a + 2b)^2 =$

15. $(11x - 2y)^2 =$

16. $(5a + 3b)^2 =$

Name____

HSA-APR.A.1

Multiplying Polynomials

Multiply each polynomial expression.

Example: $(x + 2)(3x^2 + x - 5)$

$3x^3 + x^2 - 5x$

$+ \quad 6x^2 + 2x - 10$

$3x^3 + 7x^2 \; 3x - 10$

1. $(x - 1)(2x^3 - 3x^2)$

2. $(x + 1)^3$

3. $(x + 1)(x^2 + 6x + 10)$

4. $(2x + 1)(x^3 - 6)$

5. $(x^2 - 1)(x^2 + 1)$

6. $(9x - 4)(6x^2 - x + 1)$

7. $(5x^2 + x - 8)(x - 1)$

8. $(6x^2 + 2x + 1)(x - 4)$

9. $(2x - 3)^3$

10. $(3x^2 + 1)^3$

11. $(14x + 1)(x^3 + x^2 - 7)$

12. $(^-2x + 1)(3x^3 + 2x + 1)$

13. $(11x^2 - 1)^3$

14. $(^-2x^2 + x)^3$

Name________________________________

HSA-SSE.A.2, HSA-APR.B.2

Factoring Binomials

Factor each binomial equation.

Example: $9x^2 - 4 = (3x + 2)\ (3x - 2)$

1. $4x^2 - 1 =$
2. $x^2 - 9 =$
3. $36x^2 - 9 =$
4. $100x^2 - 81 =$
5. $25x^2 - 4 =$
6. $81x^2 - 121 =$
7. $x^2 - 16 =$
8. $144x^2 - 16 =$
9. $x^2 - 25 =$
10. $625 - 16x^2 =$
11. $100 - x^2 =$
12. $x^2 - 36 =$
13. $121x^2 - 49 =$
14. $49x^2 - 16 =$

Cross out the correct answers below. Use the remaining letters to complete the statement.

$(x + 13)\ (x - 13)$ THE	$16\ (3x - 1)\ (3x + 1)$ SUM	$(x - 4)\ (x + 4)$ OFA	$(6x + 5)\ (6x - 5)$ PRO	$(25 - 4x)\ (25 + 4x)$ QUO	$(x + 1)\ (x - 1)$ DUC
$(9 + x)\ (9 - x)$ TOF	$9\ (2x - 1)\ (2x + 1)$ TIE	$(x + 7)\ (x - 7)$ THE	$(2x + 1)\ (2x - 1)$ NTA	$(9x + 1)\ (9x - 1)$ SUM	$(x + 2)\ (x - 2)$ AND
$(10 - x)\ (10 + x)$ WAS	$(5x + 3)\ (5x - 3)$ DIF	$(x - 5)\ (x + 5)$ HAS	$(8x + 1)\ (8x - 1)$ FER	$(11x - 7)\ (11x + 7)$ MAN	$(x - 6)\ (x + 6)$ NER
$(x + 18)\ (x - 18)$ ENC	$(10x - 9)\ (10x + 9)$ THA	$(x - 3)\ (x + 3)$ TIS	$(5x - 2)\ (5x + 2)$ MYP	$(7x + 11)\ (7x - 11)$ EOF	$(x + 8)\ (x - 8)$ THE
$(x + 15)\ (x - 15)$ SQU	$(9x - 11)\ (9x + 11)$ ROB	$(x + 9)\ (x - 9)$ ARE	$(3x + 2)\ (3x - 2)$ ROO	$(7x - 4)\ (7x + 4)$ LEM	$(x + 9)\ (x - 9)$ TS.

15. The factored form of the difference of the two squares is

___ _________ __ ___ ___ ___

_____________ __ ___ _______

______

Name________________________________

HSA-SSE.A.2, HSA-APR.B.2

Factoring Perfect Square Trinomials

$$x^2 - 14x + 49 = (x - 7)^2$$
$$x^2 + 6x + 9 = (x + 3)^2$$
$$a^2 \pm 2ab + b^2 = (a \pm b)^2$$

Factor. Write prime if prime.

1. $x^2 + 8x + 16 =$ ____________________
2. $x^2 - 16x + 64 =$ ____________________
3. $y^2 + 12y + 36 =$ ____________________
4. $y^2 - 10y + 25 =$ ____________________
5. $16x^2 + 8x + 1 =$ ____________________
6. $9x^2 - 6x + 1 =$ ____________________
7. $25x^2 + 10x + 1 =$ ____________________
8. $81y^2 - 90y + 25 =$ ____________________
9. $4y^2 - 20y + 25 =$ ____________________
10. $25x^2 + 60x + 36 =$ ____________________
11. $16 + 40y + 25y^2 =$ ____________________
12. $16y^2 + 24y + 9 =$ ____________________
13. $49x^2 - 14x + 1 =$ ____________________
14. $9y^2 - 30y + 25 =$ ____________________
15. $y^2 + 4y + 4 =$ ____________________

Fill in the blanks so that each expression is a trinomial square; factor.

16. $y^2 + 18x +$ ____________ = ______________________________
17. $49y^2 - 28x +$ ____________ = ______________________________
18. $9y^2 +$ ____________ $+ 36$ = ______________________________
19. $y^2 -$ ____________ $+ 121$ = ______________________________
20. ________ $y^2 + 20y + 25$ = ______________________________

Think About It

Why is the last term of a perfect square trinomial always positive?

____

____

Name________________________________

HSA-SSE.A.2, HSA-APR.A.1, HSA-APR.B.2

Un-FOIL

To factor a trinomial of the form $x^2 + bx + c$ into $(x + \quad)(x + \quad)$, undo the FOIL steps. Find two numbers that have a product equal to c and a sum equal to b. Write the two numbers in the spaces.

Example: Factor (un-FOIL) $x^2 - 7x + 12$

product = 12
sum = $^-7$
two numbers = $^-3$ and $^-4$

Therefore, $x^2 - 7x + 12 = (x + {}^-3)(x + {}^-4)$ or $(x - 3)(x - 4)$

Factor each of the following.

1. $x^2 + 6x + 8$
   product =
   sum =
   two numbers =
   $(x + \quad)(x + \quad)$

2. $x^2 + 5x + 6$
   product =
   sum =
   two numbers =
   $(x + \quad)(x + \quad)$

3. $x^2 - 9x + 14$
   product =
   sum =
   two numbers =
   $(x + \quad)(x + \quad)$

4. $x^2 + 16x - 36$
   product =
   sum =
   two numbers =
   $(x + \quad)(x + \quad)$

5. $x^2 - 8x + 15$
   product =
   sum =
   two numbers =
   $(x + \quad)(x + \quad)$

6. $x^2 - 4x - 32$
   product =
   sum =
   two numbers =
   $(x + \quad)(x + \quad)$

7. $x^2 - x - 6$
   product =
   sum =
   two numbers =
   $(x + \quad)(x + \quad)$

8. $x^2 + 3x - 18$
   product =
   sum =
   two numbers =
   $(x + \quad)(x + \quad)$

9. $x^2 + 7x - 18$
   product =
   sum =
   two numbers =
   $(x + \quad)(x + \quad)$

10. $x^2 + x - 56$
    product =
    sum =
    two numbers =
    $(x + \quad)(x + \quad)$

11. $x^2 - 22x - 75$
    product =
    sum =
    two numbers =
    $(x + \quad)(x + \quad)$

12. $x^2 - 3x - 40$
    product =
    sum =
    two numbers =
    $(x + \quad)(x + \quad)$

Name________________________________ HSA-SSE.A.2, HSA-APR.A.1, HSA-APR.B.2

Factoring Trinomials: $x^2 + bx + c$

$k^2 - k - 20 = (k)^2 + (4 + {}^-5)\,k + (4)\,({}^-5) = (k + 4)\,(k - 5)$

Factor. Write prime if prime.

1. $x^2 + 7x + 12 =$
2. $m^2 + 10m + 21 =$
3. $y^2 - 7y - 8 =$
4. $x^2 - 6x + 5 =$
5. $x^2 + 4x - 32 =$
6. $x^2 - 2x - 15 =$
7. $x^2 - 6x + 8 =$
8. $y^2 + 9y + 18 =$
9. $3 - 4t + t^2 =$
10. $v^2 + 12v + 20 =$
11. $51 - 20k + k^2 =$
12. $a^2 - 14ab + 24b^2 =$
13. $y^2 + 6y - 72 =$
14. $x^2 - 11xy - 60y^2 =$
15. $15r^2 + 2rs - s^2 =$
16. $3x^2 + 21xy - 54y^2 =$
    Hint: Check for GCF.
17. $x^2 - 5xy - 6y^2 =$
18. $x^2 + 8xy + 12y^2 =$
19. $y^2 - 7xy + 10x^2 =$
20. $a^2 - 11ab - 60b^2 =$

Name________________________________ HSA-SSE.A.2, HSA-APR.A.1, HSA-APR.B.2

Factoring Trinomials: $ax^2 + bx + c$

$2x^2 - 5x - 3 = (2x + 1)(x - 3)$

Factor. Write prime if prime.

1. $2x^2 - 5x - 3 =$
2. $3x^2 + 10x - 8 =$
3. $2y^2 + 15y + 7 =$
4. $7a^2 - 11a + 4 =$
5. $5n^2 + 17n + 6 =$
6. $4y^2 + 8y + 3 =$
7. $3x^2 + 4x - 7 =$
8. $2x^2 + 13x + 15 =$
9. $9y^2 + 6y - 8 =$
10. $6x^2 - 7x - 20 =$
11. $2n^2 - 3n - 14 =$
12. $5n^2 + 2n + 7 =$
13. $10x^2 + 13x - 30 =$
14. $12y^2 + 7y + 1 =$
15. $2n^2 + 9n - 5 =$
16. $2x^2 + 7x + 6 =$
17. $5a^2 - 42a - 27 =$
18. $15x^2 - 28x - 32 =$
19. $8a^2 - 10a + 3 =$
20. $2y^2 - 3y - 20 =$

Name________________________________ HSA-SSE.A.2, HSA-ARP.A.1, HSA-APR.B.2

More Factoring Trinomials

Factor each trinomial equation.

Example: $x^2 - 8x + 12 = (x - 6)(x - 2)$

1. $x^2 - 12x + 36 =$

2. $x^2 + 24x + 144 =$

3. $x^2 - 16x - 36 =$

4. $x^2 - 9x - 22 =$

5. $x^2 + 18x + 32 =$

6. $x^2 - x - 56 =$

7. $6x^2 + 7x + 2 =$

8. $3x^2 + 2x - 16 =$

9. $6x^2 - 5x - 4 =$

10. $15x^2 - x - 2 =$

11. $18x^2 + 9x + 1 =$

12. $20x^2 + 13x + 2 =$

13. $5x^2 - 26x + 5 =$

14. $x^2 - 9x - 10 =$

Name________________________________ HSA-SSE.A.2, HSA-ARP.A.1, HSA-APR.B.2

Factoring Polynomials

Patterns: Sum of cubes $(a^3 + b^3) = (a + b)\ (a^2 - ab + b^2)$
Difference of cubes $(a^3 - b^3) = (a - b)\ (a^2 + ab + b^2)$

Process:

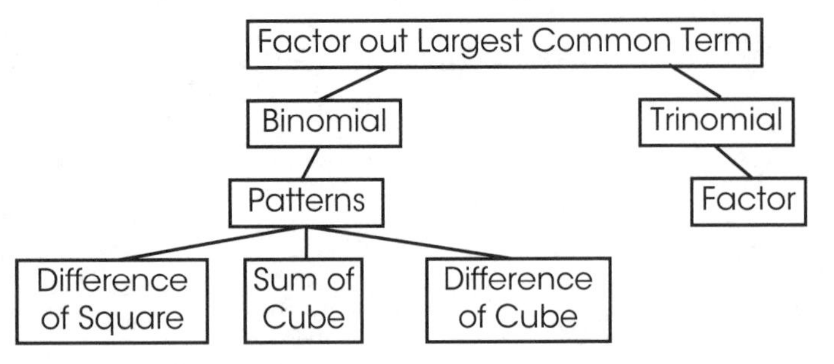

Example 1: $2x^3 + 4x + 2x$
$2x(x^2 + 2x + 1)$
$2x(x + 1)\ (x + 1)$

Example 2: $8x^3 - 125$
$a = 2x \quad b = 5$
$(2x - 5)\ [(2x)^2 + (2x)(5) + (5)^2]$
$(2x - 5)\ (4x^2 + 10x + 25)$

1. $x^3 + 5x^2 + 6x$

2. $x^3 + 1$

3. $64x^3 - 27$

4. $2x^3 - 8x^2 + 8x$

5. $x^2 - 2x - 8$

6. $x^4 + 216x$

7. $10x^2 + 5x - 5$

8. $2x^3 -7x^2 + 6x$

9. $250x^4 - 54x$

10. $40x^4 - 10x^2$

Name________________________________

HSA-APR.D.6, HSA-APR.D.7

Dividing Monomials

$$\frac{18x^6y^{10}}{-3x^3y^5} = \frac{18}{-3} \cdot \frac{x^{6-3}}{1} \cdot \frac{1}{y^{5-1}} = \frac{-6x^3}{y^4}$$

1. $\frac{m^{10}}{m^5} =$

2. $\frac{x^3y^2}{2x^2y^2} =$

3. $\frac{4ab^3}{2a^2b^2} =$

4. $\frac{27u^2v^3}{-18u^4v^5} =$

5. $\frac{13c^9d^{10}}{-26c^9d} =$

6. $\frac{3s^5t^7}{-3s^5t^7} =$

7. $\frac{-52x^3y^2z}{13xy^2} =$

8. $\frac{8xy^2}{16x^3y^5} =$

9. $\frac{5x^4}{5} =$

10. $\frac{18x^2y}{24xy} =$

11. $\frac{56s^2t^3}{4s^2t} =$

12. $\frac{48a^3bc^5}{12a^5b^3c^2} =$

13. $\frac{25x^2y}{-15xy^2} =$

14. $\frac{8m^2n^2}{12m^2n^3} =$

15. $\frac{-17c^5d^4}{-51cd^3} =$

16. $\frac{24x^2y^3z^4}{-44x^4y^3z^2} =$

Name________________________

HSA-APR.D.6, HSA-APR.D.7

Dividing a Polynomial by a Monomial

$$\frac{r^2 + 6r + 5}{r} = \frac{r^2}{r} + \frac{6r}{r} + \frac{5}{r} \text{ - } r + 6 + \frac{5}{r}$$

1. $\frac{a^2 + 2a}{a} =$

2. $\frac{14x + 35}{7} =$

3. $\frac{4y^2 + 6y}{2y} =$

4. $\frac{x^2y - xy^2}{xy} =$

5. $\frac{25u^2 - 15u - 5}{^-5} =$

6. $\frac{12x^2 - 9x^3 + 6x^4}{3x} =$

7. $\frac{m^2n^2 + m - n}{mn} =$

8. $\frac{45a^2b^4 - 60a^3b^2 - 15a^2b}{-15a^2b} =$

9. $\frac{14k^9m^3 - 4k^2m^2 + 12km^3}{2km^2} =$

10. $\frac{12v^5 - 27v^4 + 18uv^3}{3uv^3} =$

11. $\frac{2x^2 - 10xy}{2x} =$

12. $\frac{3x^3y^2 - 6x^2y^2 + 6xy^2}{3xy} =$

13. $\frac{6z^2 - 3z + 9}{3z} =$

14. $\frac{6a^2 + 42a + 72}{6a^3} =$

15. $\frac{64x^4 - 64x^3}{64x^3} =$

16. $\frac{18m^3n^4 - 12m^2n^3 + 24n^2}{6m^2n} =$

Name________________________________

HSA-APR.D.6, HSA-APR.D.7

Dividing Polynomials

$$\frac{6a^2 + 4a + 3}{3a - 1} \Longrightarrow \begin{array}{r|l} & 2a + 2 + \frac{5}{3a-1} \\ \hline 3a - 1 & 6a^2 + 4a + 3 \\ & \underline{6a^2 - 2a} \\ & \quad 6a + 3 \\ & \quad \underline{6a - 2} \\ & \quad\quad 5 \end{array}$$

1. $\frac{s^2 + 3s - 4}{4 + s} =$

   Hint: Rewrite denominator as $s + 4$.

2. $\frac{a^2 + 2a + 3}{a + 3} =$

3. $\frac{x^2 + 4}{x + 2} =$

   Hint: Write dividend as $x^2 + 0x + 4$.

4. $\frac{3c^2 + 6c + 4}{3c + 2} =$

5. $\frac{6r^2 + r - 5}{2r - 3} =$

6. $\frac{9t^2 + 1}{3t + 2} =$

7. $\frac{2u^2 - 3uv - 9v^2}{u - 3v} =$

8. $\frac{z^3 + z^2 - 3z + 9}{z + 3} =$

9. $\frac{6x^3 + 5x^2 + 9}{2x + 3} =$

10. $\frac{2y^3 + 5y^2 + 7y + 6}{y^2 + y + 2} =$

11. $x^3 - x^2 - 2x + 10 \div (x + 2) =$

12. $8x + 13x^2 + 6x^3 + 5 \div (3x + 5) =$

13. $y^3 - 2y^2 + 3 \div (y + 1) =$

14. $\frac{^-32x + 2x^3 + 42}{2x - 6} =$

Name________________________________ HSA-APR.D.6, HSA-APR.D.7

More Dividing Polynomials

Divide each polynomial expression. Check your answer.

Example: $\frac{(4x^2 - 2x + 6)}{(2x - 3)}$

$$\begin{array}{r} 2x + 2 \\ 2x - 3 \overline{\smash{)}\, 4x^2 - 2x + 6} \\ -\underline{4x^2 + 6x} \\ 4x + 6 \\ \underline{-4x + 6} \\ 12 \text{ remainder} \end{array}$$

$(2x + 2) + \frac{12}{2x - 3}$

Check: $(2x + 2)(2x - 3) + 12$
$4x^2 - 2x - 6 + 12 = 4x^2 - 2x + 6$

1. $(x^3 - 1) \div (x^2 - 1)$

2. $\frac{(2x^2 - 5x - 3)}{x - 3}$

3. $\frac{(x^2 - 3x - 7)}{x + 2}$

4. $(x^3 - 6) \div (x - 1)$

5. $(x^3 - 6x^2 + 1) \div (x + 2)$

6. $(5x^2 - 34x - 7) \div (x - 7)$

7. $\frac{(x^4 - 3x^3 - 5x - 6)}{(x + 2)}$

8. $\frac{(6x^2 - x - 7)}{(3x + 1)}$

Name________________________________ HSA-APR.D.6, HSA-APR.D.7

Synthetic Division

Synthetic division is a short-hand method of dividing polynomials when the divisor is a binomial expression with a one as the coefficient of the variable. It serves as a useful tool in factoring and graphing polynomials. At first glance, the process seems involved. However, with practice, synthetic division is the fastest way to divide two polynomials.

The Process	Example
1. Write the polynomial in descending order.	$(-4x^2 + 3x^3 - 2 - 3x) \div (x - 3)$ $(3x^3 - 4x^2 - 3x - 2)$
2. Write the coefficients in a row.	$3 \quad -4 \quad -3 \quad -2$
3. Write the opposite of the constant of the divisor in front of the row of coefficients.	$\begin{array}{r\|rrrr} 3 & 3 & -4 & -3 & -2 \end{array}$
4. Bring the first coefficient straight down.	$\begin{array}{r\|rrr\|r} 3 & 3 & -4 & -3 & -2 \\ & & & & \\ \hline & 3 & & & \end{array}$
5. Multiply the first coefficient by the divisor and place the product under the second coefficient.	$\begin{array}{r\|rrr\|r} 3 & 3 & -4 & -3 & -2 \\ & & 9 & & \\ \hline & 3 & & & \end{array}$
6. Add the second coefficient to the product from step #5.	$\begin{array}{r\|rrr\|r} 3 & 3 & -4 & -3 & -2 \\ & & 9 & & \\ \hline & 3 & 5 & & \end{array}$
7. Multiply the sum from step #6 by the divisor and place under the third coefficient.	$\begin{array}{r\|rrr\|r} 3 & 3 & -4 & -3 & -2 \\ & & 9 & 15 & \\ \hline & 3 & 5 & & \end{array}$
8. Add the third coefficient to the product from step #7.	$\begin{array}{r\|rrr\|r} 3 & 3 & -4 & -3 & -2 \\ & & 9 & 15 & \\ \hline & 3 & 5 & 12 & \end{array}$
9. Multiply the sum from step #8 by the divisor and place it under the fourth coefficient.	$\begin{array}{r\|rrr\|r} 3 & 3 & -4 & -3 & -2 \\ & & 9 & 15 & 36 \\ \hline & 3 & 5 & 12 & \end{array}$
10. Add the fourth coefficient to the product from step #9.	$\begin{array}{r\|rrr\|r} 3 & 3 & -4 & -3 & -2 \\ & & 9 & 15 & 36 \\ \hline & 3 & 5 & 12 & 34 \end{array}$
11. Each coefficient in the last row corresponds to a new polynomial with the degree of the polynomial being decreased by one.	$3x^2 + 5x + 12$
12. If the sum from step #10 is any expression other than zero, write it over the original divisor.	$\dfrac{3x^2 + 5x + 12 + 34}{x - 3}$

Name________________________________

HSA-APR.D.6, HSA-APR.D.7

More Synthetic Division

Use synthetic division to divide the polynomials.

1. $(16 - 8x - 7x^2 + 2x^3) \div (x - 4)$

2. $(2x^2 + 3 + 5x) \div (x + 1)$

3. $(3x + x^2 - 18) \div (x + 3)$

4. $(2x^2 - 4x + 3) \div (x - 3)$

5. $(x - x^2 + 8 + x^3) \div (x - 1)$

6. $(-x^2 + 2x - 4 + 3x^3) \div (x - 2)$

Name________________________________

HSA-APR.B.2, HSA-APR.D.6, HSA.APR.D.7

Factoring Using Synthetic Division

Sometimes, when factoring polynomials, nothing seems to work. For example, there may not be a recognizable pattern or trinomial. In such cases, we can use synthetic division to reduce the polynomial to a trinomial or binomial, then look for patterns which are familiar.

In the following example, there are no familiar patterns or trinomials. The goal is to reduce the polynomial using synthetic division by trial and error and end up with a remainder of zero. If the remainer is not zero, choose a different factor and try again.

Hint: Try factors of the constant term for your trial and error.

Example: Factor $x^3 - 2x^2 - 7x - 4$ using synthetic division.

Start by assuming that $(x - 2)$ is a factor of this polynomial.

2	1	-2	-7	-4
		2	0	-14
	1	0	-7	-18

Since the remainder is not zero, $(x - 2)$ is not a factor of this polynomial.

This time, try $(x - 4)$ as a factor.

4	1	-2	-7	-4
		4	8	4
	1	2	1	0

Since the remainder is zero, $(x - 4)$ is a factor of the polynomial $x^3 - 2x^2 - 7x - 4$.

Now, rewrite the bottom line as a trinomial. $x^2 + 2x + 1$

Multiply the new trinomial by the factor $(x - 4)$. The problem at this point is $(x - 4)(x^2 + 2x + 1)$.

Factor the familiar trinomial. The final factors can be written in simplest terms.
$(x - 4)(x + 1)(x + 1)$

Name________________________________

HSA-APR.B.2, HSA-APR.D.6, HSA-APR.D.7

More Factoring Using Synthetic Division

Use synthetic division to factor the polynomials. **Hint:** Try factors of the constant term for your trial and error.

1. $2x^3 - 3x^2 - 3x + 2$

2. $x^3 - x^2 - x + 1$

3. $x^3 + 7x^2 + 7x - 15$

4. $x^3 - x^2 - 10x - 8$

5. $2x^3 - 3x^2 - 2x + 3$

6. $3x^3 - 13x^2 - 11x + 5$

Name________________________________ HSA-APR.B.3

Graphing Polynomials

Graph the polynomials using x-intercepts.

Process: 1. Factor the polynomial.
2. Solve for the x-intercept.
3. Graph the x-intercept.
4. Determine the general shape.
5. Graph.

Example: $y = 2x^3 + 4x^2 + 2x$

step 1: $= 2x(x^2 + 2x + 1)$
$= 2x(x + 1)(x + 1)$

step 3:

y
x
(-1, 0)
(0, 0)

step 2: $2x = 0 \rightarrow x = 0$
$x + 1 = 0 \rightarrow x = {}^-1$
$x + 1 = 0 \rightarrow x = {}^-1$

step 4: The shape of the graph is determined by the term with the highest power of x.
Plug in two x values; one positive and one negative.

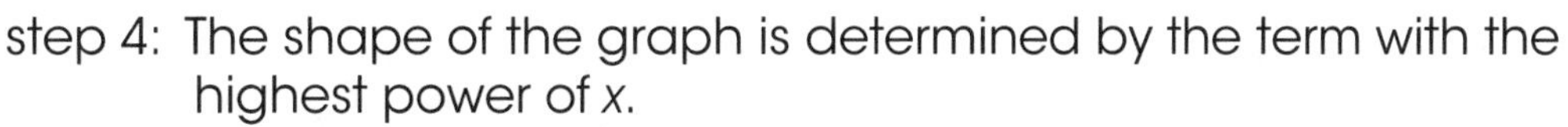

$2x^3$
$2({}^-4)^3 = {}^-128$
$2(4)^3 = 128$

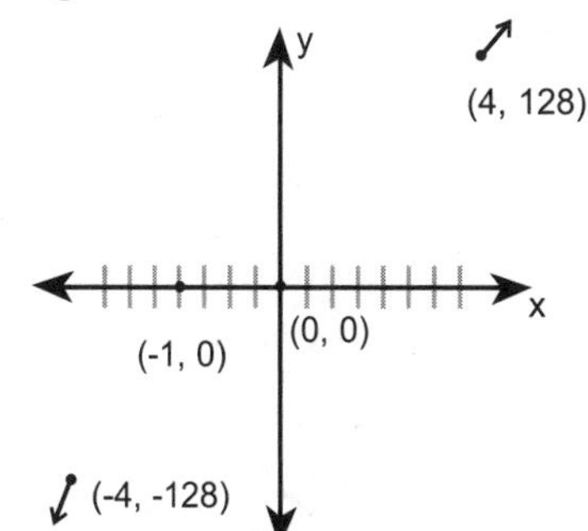

step 5: Since the same solution for the x-intercept appears twice (in this example, $x = {}^-1$), the graph hits the x-axis at this point and moves away, not through the -axis.

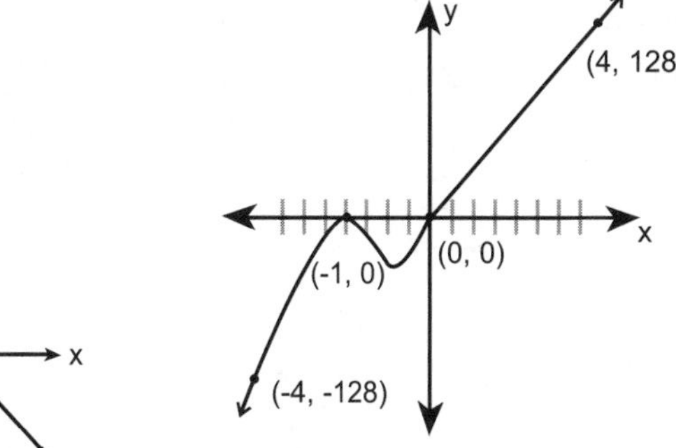

Note: If the coefficient on the x^3 term were negative, the shape of the graph would also change.

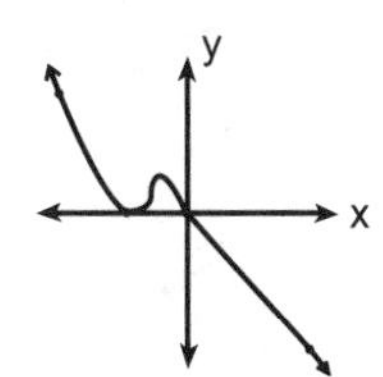

Name________________________________ HSA-APR.B.3

More Graphing Polynomials

Graph the polynomials using the *x*-intercept.

1. $2x^3 - 7x^2 + 6x = y$

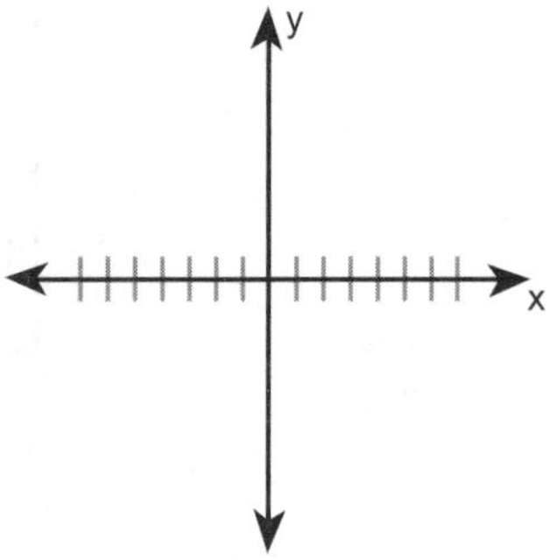

2. $-x^3 + 5x^2 - 6x = y$

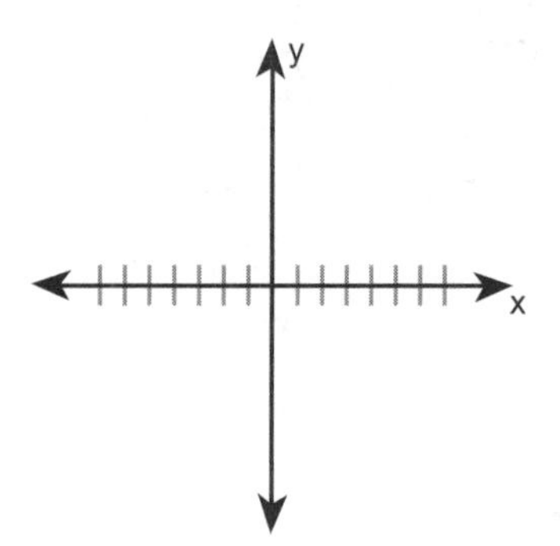

3. $^{-}4x^3 - 12x^2 + x + 3 = y$

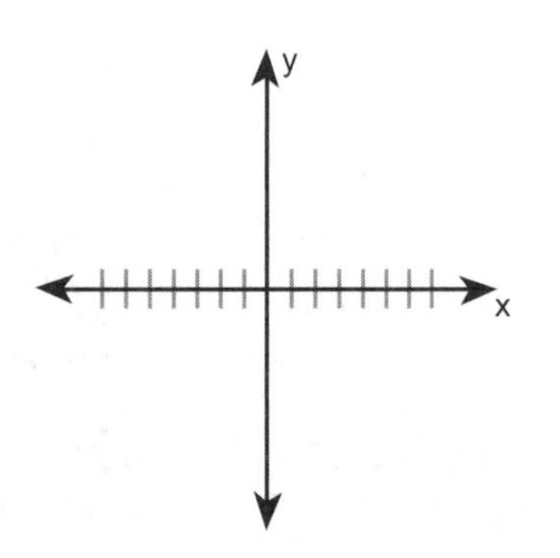

4. $x^3 + x^2 - x - 1 = y$

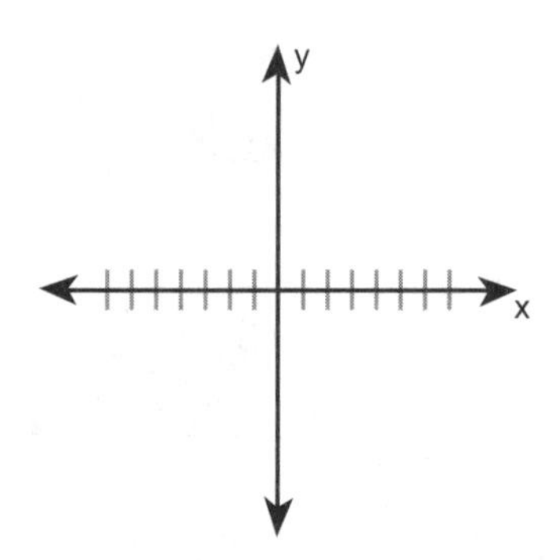

5.$40x^4 - 10x^2 = y$

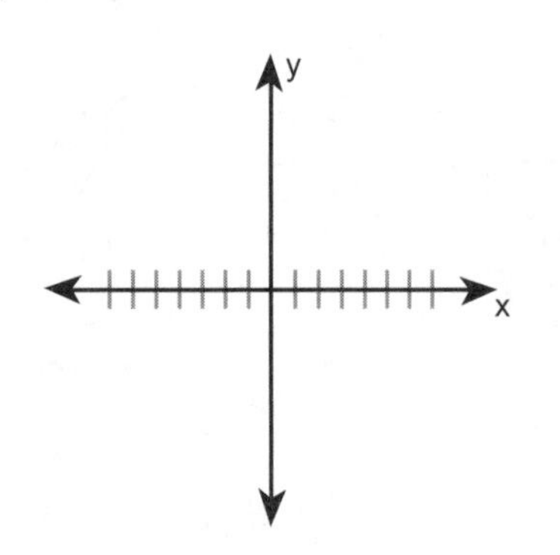

6. $x^4 - 9x^3 + 20x^2 = y$

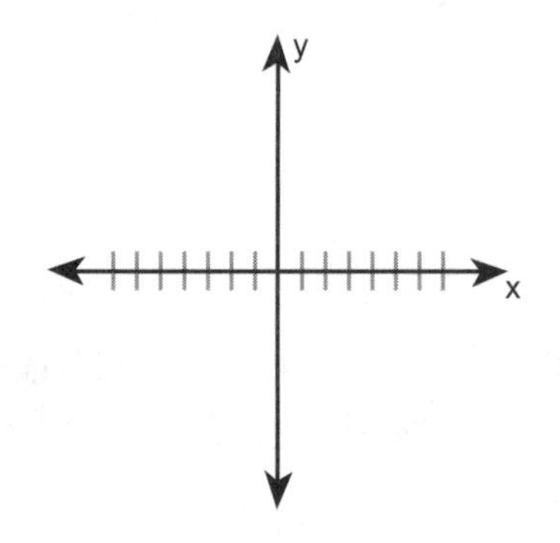

Name______________________________

HSA-REI.A.2

Finding Real Roots I

Find all real roots for each equation.

Example: $\frac{12x^4}{12} = \frac{15552}{12}$

$x^4 = 1296$

$x = 6$ or $^-6$

Note: An even root gives two values.

1. $\sqrt[3]{x} + 2 = 9$

2. $2x^2 = 200$

3. $^-6x^7 = {}^-768$

4. $x^4 = 625$

5. $-4\sqrt[3]{x} = -20$

6. $2\sqrt{x} = 10$

7. $^-10\sqrt{x} = {}^-1000$

8. $13x^5 = 3159$

9. $^-14\sqrt[3]{x} = 378$

10. $^-11x^3 = 704$

Calculator Challenge

11. $\sqrt[4]{x} = 81$

Above each problem number, place the letter associated with its answer from the Answer Bank.

Non-real roots of an equation are the

__ __ __ __ __ __ __ __ __   __ __ __ __ __ .
2 3 4 7 2 1 4 5 8   5 6 6 9 10

Answer Bank

A.	C.	D.	E.	G.	H.	I.	K.	L.	M.
±5	±36	±9	100	10,000	1,000	±10	⁻81	16	2
N.	**O.**	**P.**	**R.**	**S.**	**T.**	**U.**	**V.**	**W.**	**Y.**
343	25	8	125	⁻4	⁻19,683	64	±12	⁻27	3

Name________________________________

HSA-REI.A.2

Finding Real Roots II

Find all real roots of each equation.

Example: $\sqrt{3x+1} = \frac{60}{3}$

$$\sqrt{3x+1}^2 = (20)^2$$
$$3x + 1 = 400$$
$$3x + 1 - 1 = 400 - 1$$
$$\frac{3x}{3} = \frac{399}{3}$$
$$x = 133$$

1. $^-9(2x + 6)^5 = 288$

2. $5\sqrt[3]{2x+1} = {}^-35$

3. $3(x - 1)^2 = 432$

4. $\frac{1}{2}(x - 2)^2 = 8$

5. $3(2x - 5)^4 = 19683$

6. $(5x - 1)^4 = 2401$

7. $^-6\sqrt{10x - 18} = {}^-294$

8. $11(12x - 1)^3 = 2376$

9. $8(4x + 3)^{\frac{1}{2}} = 648$

10. $7(x - 15)^{\frac{3}{5}} = 2401$

11. $^-4(13x - 23)^{\frac{3}{4}} = {}^-32$

Name________________________________

HSA-REI.B.3, HSA-REI.B.4b

Solving Quadratic Equations

$(x - 5)^2 = 36$

$\sqrt{(x-5)^2} = \sqrt{36}$

$x - 5 = \pm 6$

$x = 11, {}^-1$

1. $x^2 = 25$

2. $(x - 2)^2 = 9$

3. $2y^2 = 32$

4. $x^2 - 49 = 0$

5. $3a^2 - 1 = 11$

6. $(2x - 5)^2 = 49$

7. $(x + 1)^2 = 4$

8. $(x + 17)^2 = 49$

9. $(x + 3)^2 = 0$

10. $4(y + 5)^2 = 4$

11. $(2x - 6)^2 = 16$

12. $3(2y + 7)^2 = 27$

Name________________________________ HSA-REI.B.3, HSA-REI.B.4b

The Quadratic Formula

$$x = \frac{-b \pm \sqrt{b^2 - 4ac}}{2a}$$

$$3x^2 - 5x - 4 = 0$$

$a = 3$, $b = {}^-5$, $c = {}^-4$ ⟹ $\frac{5 \pm 2\sqrt{5 - 4(3)({}^-4)}}{6} = \frac{5 \pm 73}{6}$

Solve using the quadratic formula.

1. $x^2 - 2x - 8 = 0$

2. $y^2 + 11y + 10 = 0$

3. $x^2 + 2x - 4 = 0$

4. $y^2 + 5y - 7 = 0$

5. $2x^2 - 3x - 5 = 0$

6. $2y^2 + 4y = 1$

7. $7x^2 + 4x - 5 = 0$

8. $3x^2 + 10x + 5 = 0$

9. $2y^2 = 3y + 4$

10. $8x^2 + 7x - 2 = 0$

11. $x^2 = 4x$

12. $\frac{3}{x - 1} - 4 = \frac{1}{x + 1}$

Name________________________________ HSA-REI.D.10, HSA-REI.D.11, HSF-LE.A.2

Graphing Quadratics: $y = ax^2$

Graph each quadratic equation using a table of values. Charts may vary.

Example: $y = x^2$

x	y
-2	**4**
-1	**1**
0	**0**
1	**1**
2	**4**

1. $y = 2x^2$

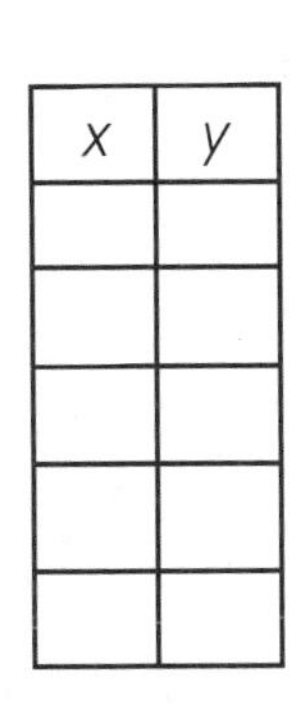

x	y

2. $y = -x^2$

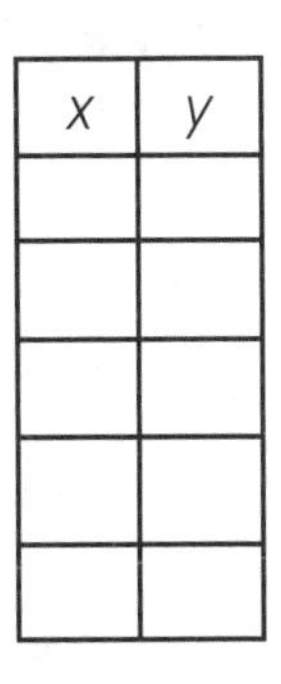

x	y

3. $y = {}^{-}3x^2$

x	y

4. $y = \frac{1}{3}x^2$

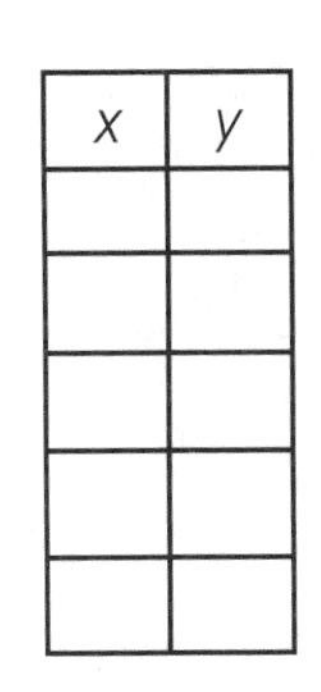

x	y

5. $y = \frac{-1}{2}x^2$

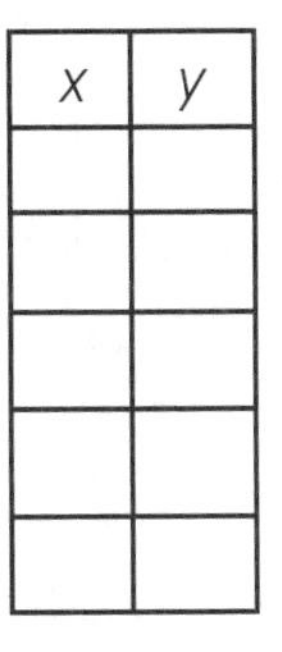

x	y

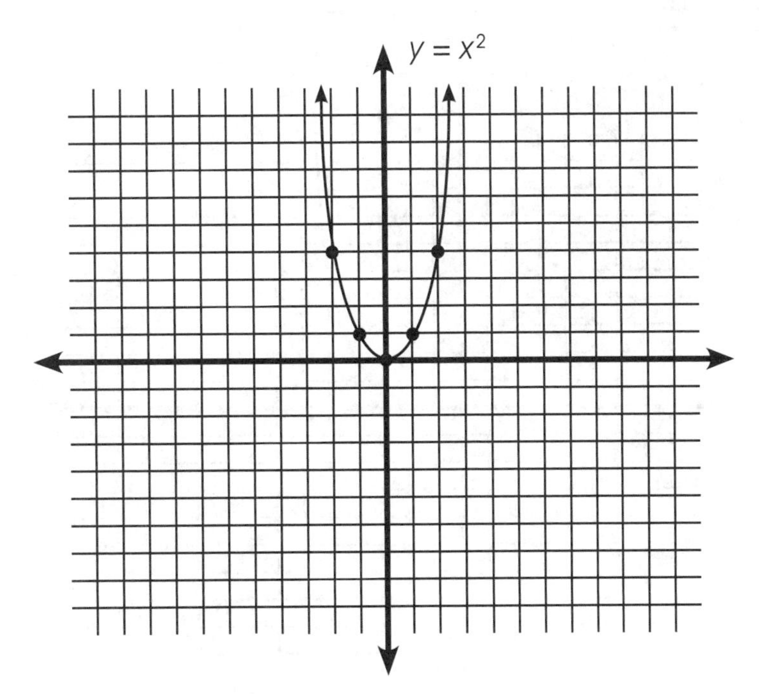

How do the constants a and $-a$ alter the graph?

Name________________________________ HSA-REI.D.10, HSA-REI.D.11, HSF-LE.A.2

Graphing Quadratics: $y = ax^2 + c$

Graph each quadratic equation.

Example: $y = x^2 + 1$

x	y
-2	5
-1	2
0	1
1	2
2	5

1. $y = x^2 - 2$

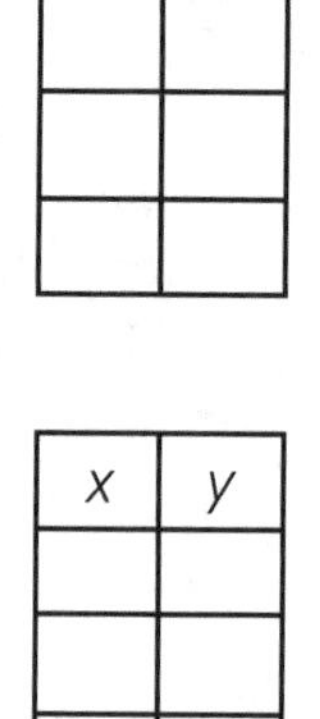

2. $y = 2x^2$

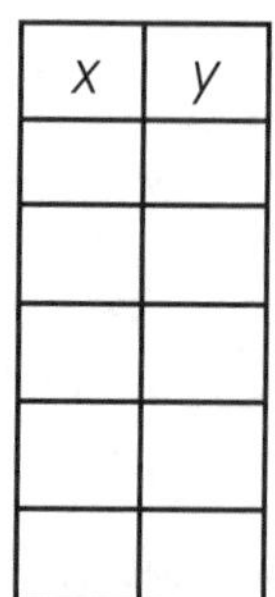

3. $y = \frac{1}{2}x^2 - 9$

x	y

4. $y = {}^{-}3x^2 + 6$

x	y

5. $y = \frac{2}{3}x^2 - 3$

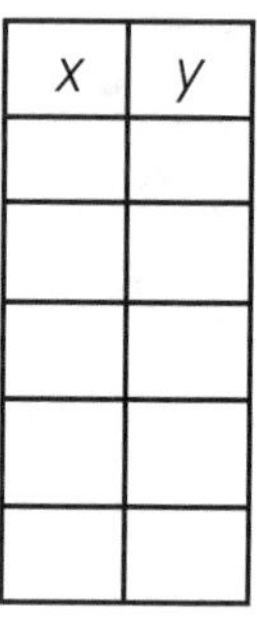

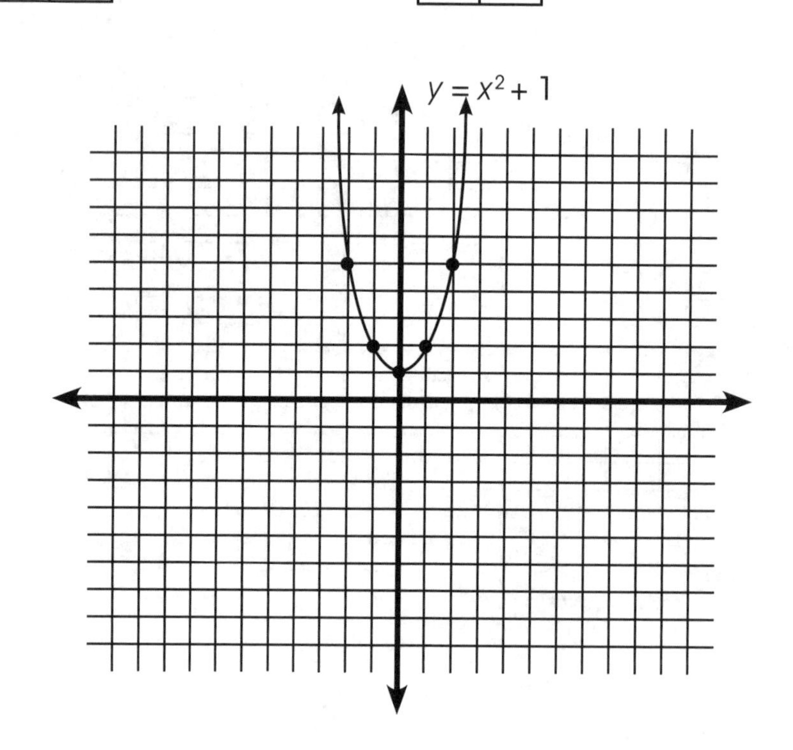

How do the constants c and $-c$ alter the graph?

Name______________________________

HSA-REI.D.10, HSA-REI.D.11, HSF-LE.A.2

Graphing Quadratics: $y = (x - b)^2 + c$

Graph each quadratic equation. Chart may vary.

Example: $y = (x - 2)^2$

x	**y**
0	**4**
1	**1**
2	**0**
3	**1**
4	**4**

1. $y = 2(x + 3)^2$

x	y

2. $y = \frac{1}{3}(x - 5)^2$

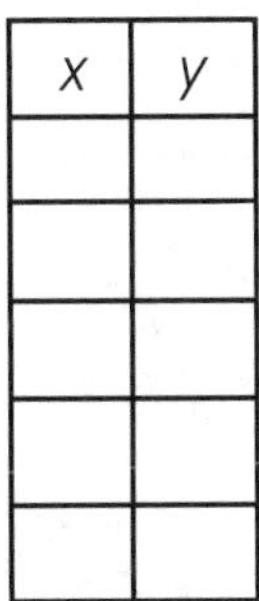

x	y

3. $y = {}^-2(x - 7)^2$

x	y

4. $y = \frac{1}{2}(x + 3)^2 - 1$

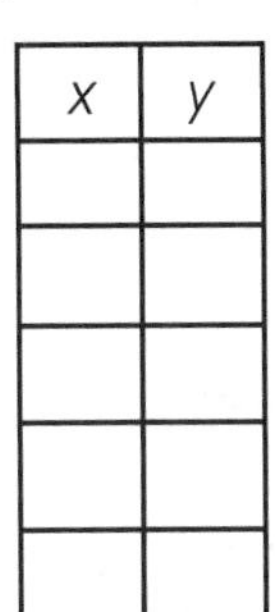

x	y

5. $y = -(x + 8)^2 + 1$

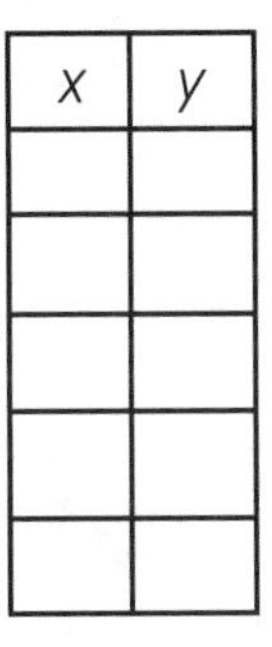

x	y

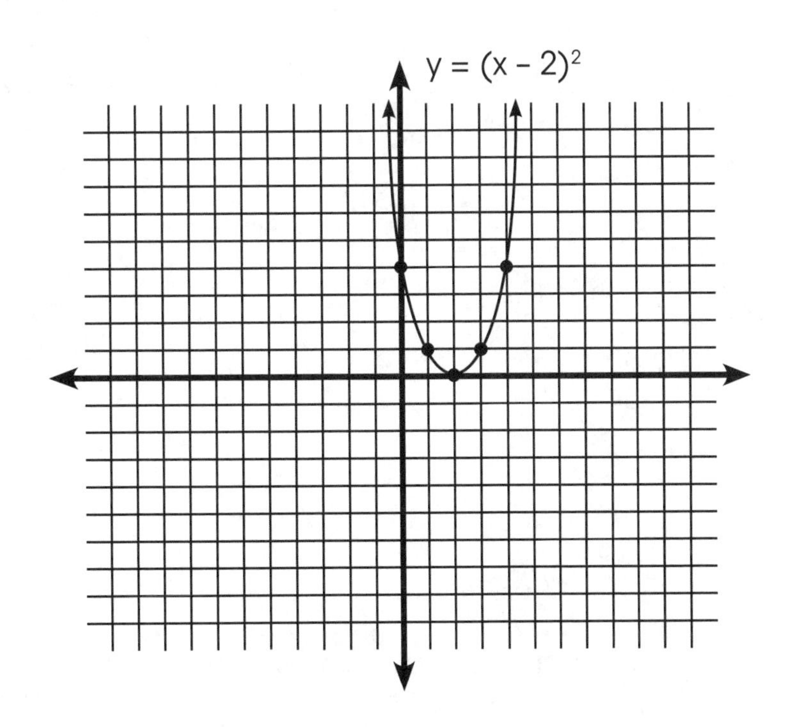

How do the constants *b* and -*b* alter the graph?

Name____________________________________

HSA-SSE.A.3b, HSF-IF.C.7a

Finding the Vertex of Quadratic Equations

Find the vertex of each quadratic using the following equation:

If $y = ax^2 + bx + c$, then vertex = $\left(\frac{^-b}{2a}, \frac{^-b^2}{4a} + c\right)$

Example: $y = x^2 + 2x + 1$

$a = 1$
$b = 2$
$c = 1$
$\left(\frac{^-2}{2(1)}, \frac{^-(2)^2}{4(1)} + (1)\right) = (-1, 0)$

1. $y = x^2 - 2x + 1$

2. $y = 8x^2 - 16x + 1$

3. $y = {^-3}x^2 + 6x - 1$

4. $y = x^2 - 9$

5. $y = 2x^2 + 1$

6. $y = x^2$

7. $y = {^-10}x^2$

8. $y = 2x^2 + 6x + 0$

9. $y = {^-4}x^2 - 7$

10. Once you have found the x-coordinate by using $\frac{^-b}{2a}$, how can you find the y-coordinate if you don't remember $\left(\frac{^-b^2}{4a}\right) + c$?

____

Name_______________________________ HSF-IF.C.7a

Finding *x*-Intercepts

Find the *x*-intercepts using the following quadratic formula:

If $y = ax^2 + bx + c$, then $x = \dfrac{^-b \pm \sqrt{b^2 - 4ac}}{2a}$

Example: $y = x^2 + 2x - 1$

$a = 1 \quad b = 2 \quad c = ^-1$

$$x = \frac{^-2 \pm \sqrt{2^2 - 4(1)(^-1)}}{2(1)} = \frac{^-2 \pm \sqrt{8}}{2}$$

$$= \frac{^-2 \pm 2\sqrt{2}}{2} = ^-1 \pm \sqrt{2}$$

1. $y = 3x^2 - 14x + 1$

2. $y = 2x^2 - x - 1$

3. $y = ^-3x^2 - 6x + 1$

4. $y = 3x^2 + x$

5. $y = ^-7x^2 - 14x$

6. $y = 4x^2 - 9$

7. $y = x^2 - 2$

8. $y = (x - 3)\ (x + 2)$

9. $y = (x)\ (2x - ^-6)$

Name____ HSF-IF.B.4, HSF-IF.C.7a

Quick Graphs of Quadratic Equations

Create a quick graph using the *x*-intercept and the vertex.

Example: $y = x^2 - 5x - 14$

x-intercept:

$(x - 7)(x + 2)$

$x - 7 = 0 \quad x = 7$

$x + 2 = 0 \quad x = {}^{-}2$

vertex:

$\left(\frac{{}^{-}b}{2a}, \frac{{}^{-}b^2}{4a} + c\right)$

$\left(\frac{{}^{-}(-5)}{(2)(1)}, \frac{{}^{-}(-5)^2}{(4)(1)} + {}^{-}14\right)$

$\left(\frac{5}{2}, \frac{81}{4}\right)$

$a = 1$

$b = {}^{-}5$

$c = {}^{-}14$

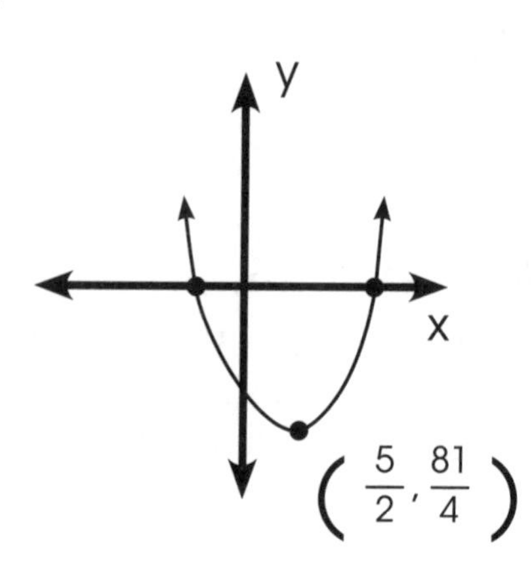

$\left(\frac{5}{2}, \frac{81}{4}\right)$

1. $y = x^2 - 8x + 12$

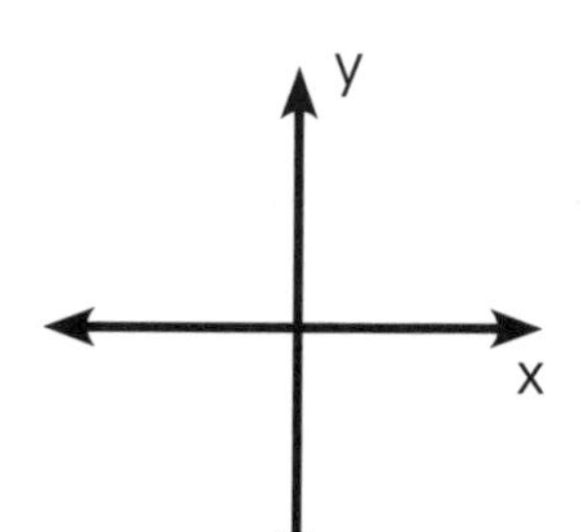

2. $y = 6x^2 - 10x - 4$

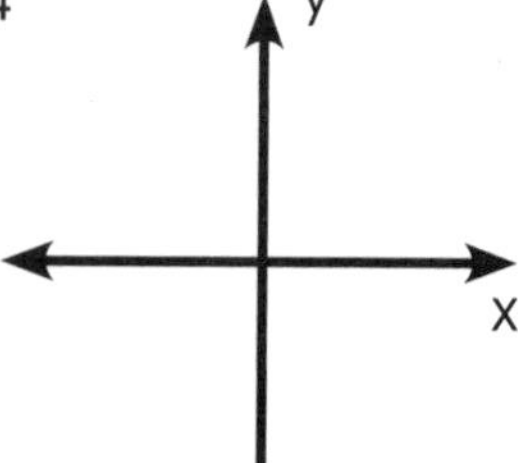

3. $y = {}^{-}15x^2 + x + 2$

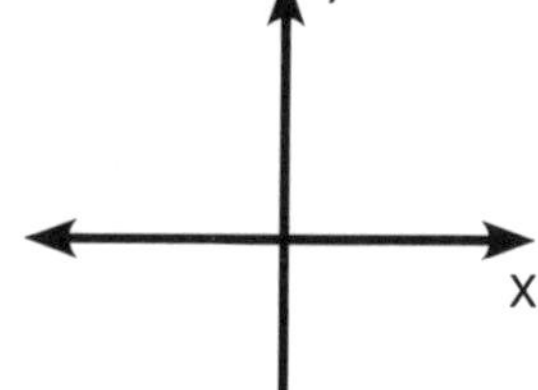

4. $y = x^2 - 16$

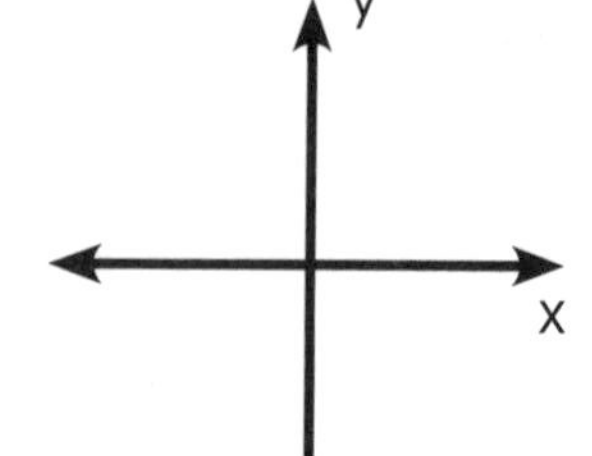

5. $y = (x - 2)(x + 3)$

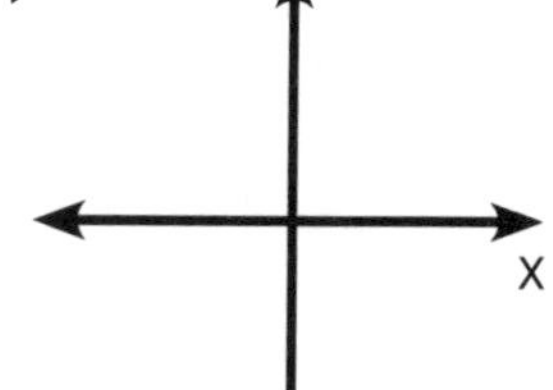

6. $y = (x - 5)(x + 5)$

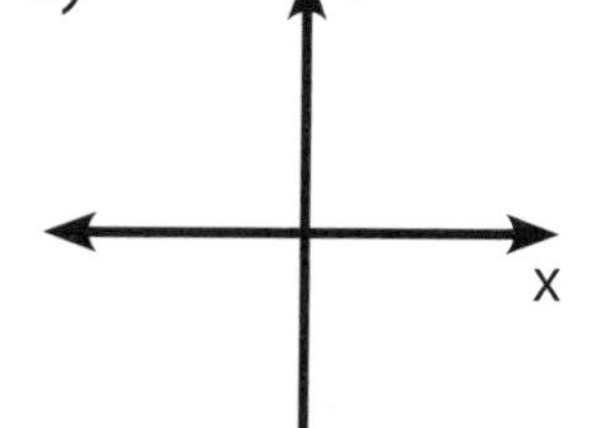

7. $y = -x^2 - 5x - 6$

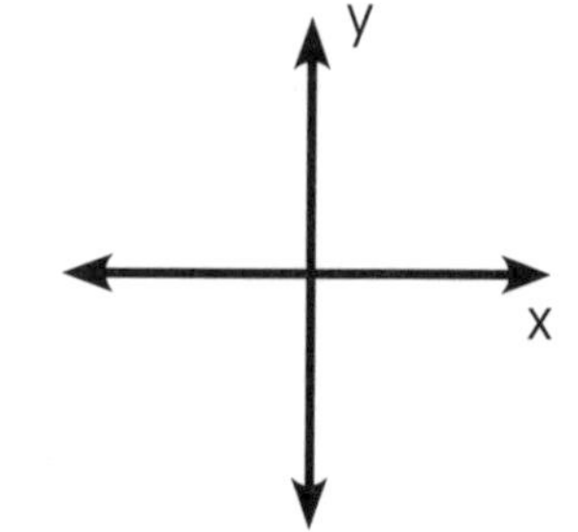

8. $y = 3x^2$

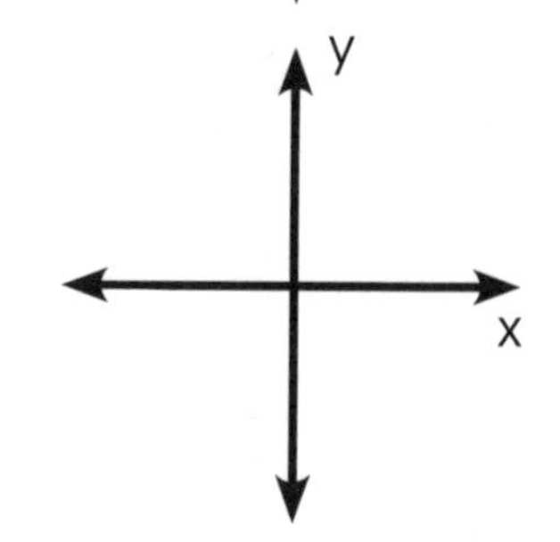

Name________________________________ HSF-IF.C.7a

Graphing Quadratic Equations

Graph each quadratic equation on a piece of graph paper.

1. $y = {}^{-}5x^2 + 1$

2. $y = (x - 1)(x + 5)$

3. $y = {}^{-}4x^2$

4. $y = 3x^2 - x$

5. $y = x^2 + 6x + 8$

6. $y = {}^{-}3x^2 - 6x + 1$

7. $y = {}^{-}4x^2 - 7$

8. $y = x^2 + 2x - 3$

9. $y = 2x^2 + 1$

10. $y = -x^2 - x - 1$

Name________________________________

Just for Fun

Make your own matrix.

Billy Brown, Willy White, Bobby Blue, and George Green all live on the same street. Their houses are painted brown, white, blue, and green, but no boy lives in a house that matches his last name. Also, each boy has a pet, and its name does not begin with the same letter as its owner's name. Also, you must find out the location of each house — is it the first, second, third, or fourth on the block?

1. George Green owns the bear.
2. Willy White owns the bull.
3. The white house is the last one on the street.
4. Neither the bear nor the bull lives next to the first house.
5. Bobby Blue's house is not green.
6. The boy who owns the whale lives in the green house.
7. The gorilla lives in the first house, which is brown.

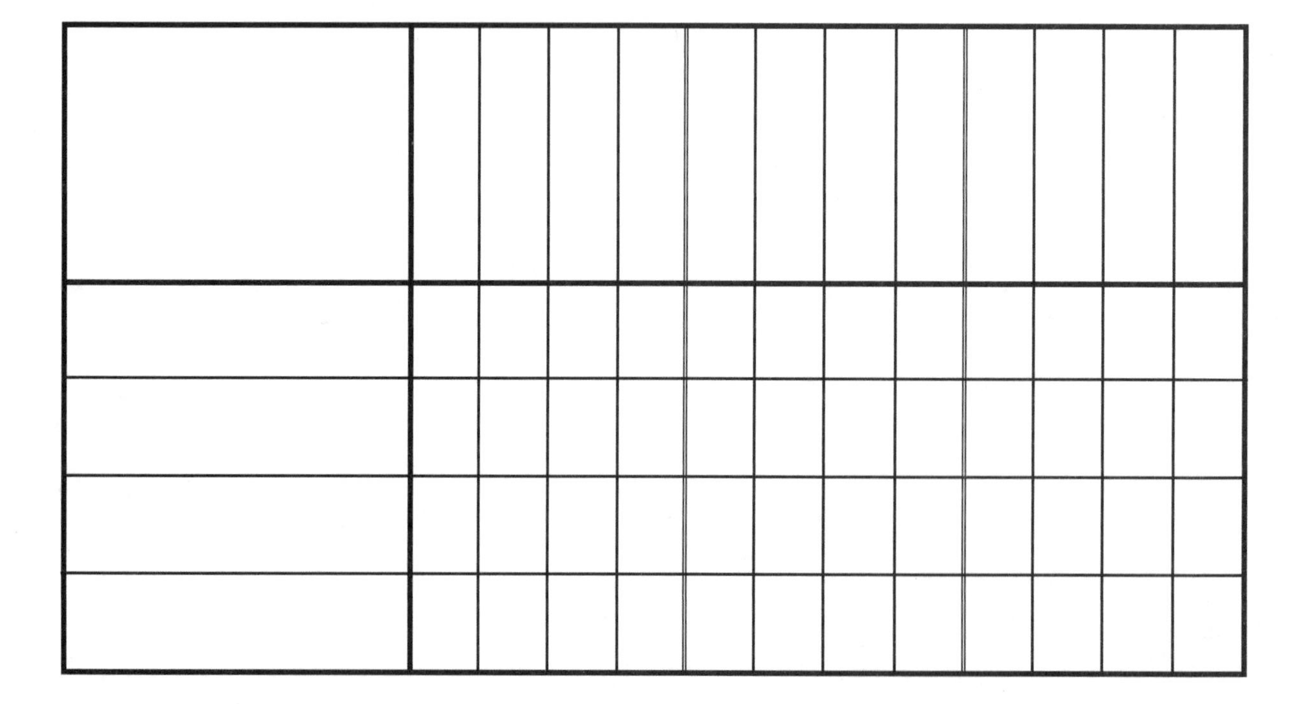

Name________________________________

HSF-LE.A.4

Manipulating Common Logs (Base 10)

$y = \log_b x$ where b = base
Common logarithm $b = 10$.
When no base is given, assume base 10.
$y = \log_{10} x$ is equivalent to $10^y = x$

Solve without using a calculator.

Example: $\log_{10} 100 = y$
$10^y = 100$
$y = 2$

1. $\log 1000$

2. $\log \sqrt[5]{10}$

3. $\log \sqrt[3]{10^2}$

4. $\log 0.1$

5. $\log 0.0001$

6. $\log \sqrt[4]{10}$

7. $\log \sqrt{10}$

8. $\log 10^6$

9. $\log 1$

10. $\log 10{,}000$

Name____________________________________

HSF-LE.A.4

Converting from Logarithmic to Exponential Form

Convert each equation from logarithmic form to exponential form or from exponential to logarithmic. $y = \log_b x \leftrightarrow b^y = x$

Example: $\log_{11} 121 = 2$
$11^2 = 121$

1. $5^3 = 125$

2. $10^6 = 1{,}000{,}000$

3. $\log_{10} 1 = 0$

4. $\log_3 \frac{1}{243} = {}^{-}5$

5. $7^5 = 16,\ 807$

6. $y = \log x$

7. $12^x = 87$

8. $y = \log_{15} 30$

9. $y = \log_Q X$

10. $y = \log_{180} B$

11. $10^y = x$

12. $\log_b 64 = 3$

13. $\log_x 5 = 10$

14. $7^x = 343$

Name________________________________

HSF-LE.A.2, HSF-LE.A.4

Graphing Logarithms

Complete the table of values and graph each function on graph paper.

Example:
$y = \log_{10} x$
Convert to exponential form.
$10^y = x$
Choose y values that are both positive and negative.

x	y
$\frac{1}{100}$	$^-2$
$\frac{1}{10}$	$^-1$
1	0
10	1
100	2
1000	3

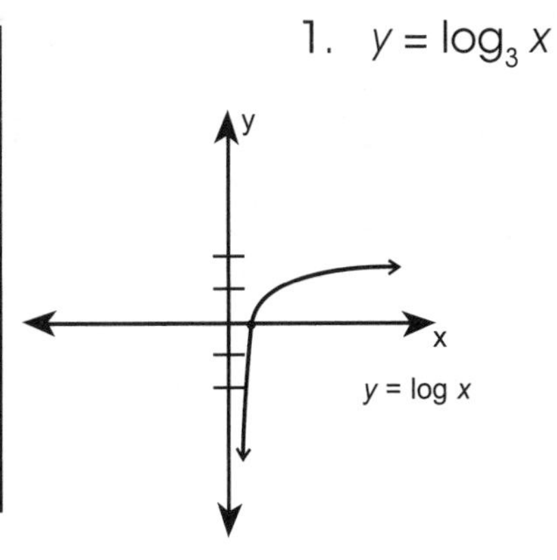

1. $y = \log_3 x$

x	y

2. $y = \log_5 x$

x	y

3. $y = \log_7 x$

x	y

4. $y = \log_2 x$

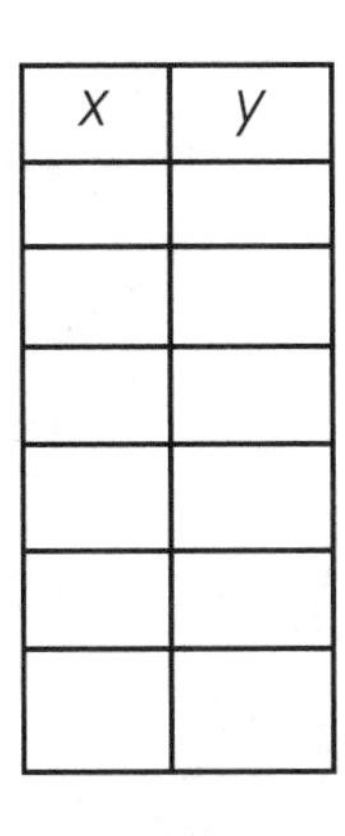

x	y

5. $y = \log_4 x$

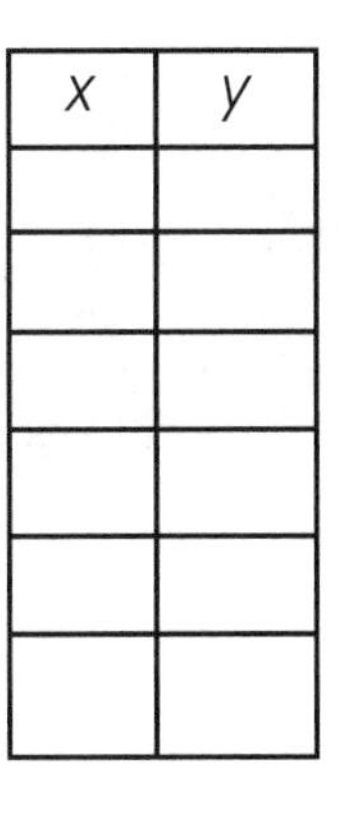

x	y

6. $y = \log_{11} x$

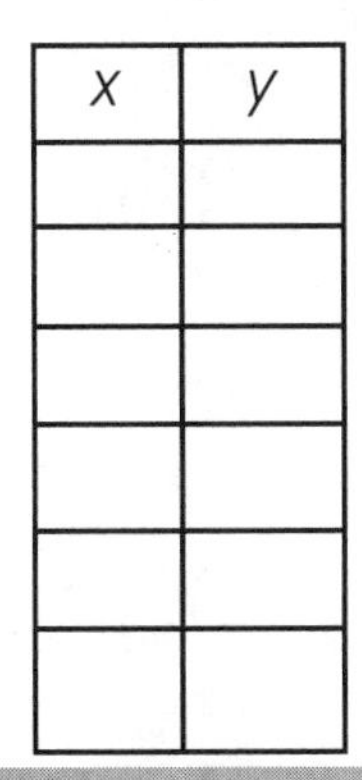

x	y

7. $y = \log_{15} x$

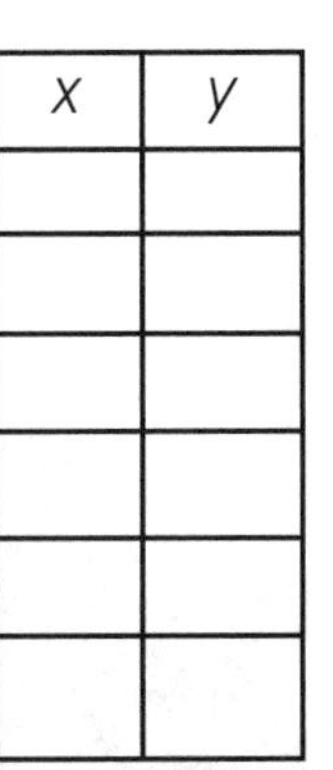

x	y

Name________________________________

HSF-BF.B.5

Simplifying Logarithms

product property: $\log_b (m \cdot n) = \log_b m + \log_b n$

quotient property: $\log_b \left(\frac{m}{n}\right) = \log_b m - \log_b n$

power property: $\log_b (m^n) = n \cdot \log_b m$

Simplify.

Example 1: $\log_5 6 + \log_5 8 = \log_5 (6 \cdot 8) = \log_5 48$

Example 2: $\log_7 9 - \log_7 3 = \log_7 \frac{9}{3} = \log_7 3$

Example 3: $\log_{12} 6^3 = 3 \log_{12} 6$

1. $\log_9 4 + \log_9 6$

2. $\log_{12} 12 + \log_{12} 11$

3. $\log_{16} 36 - \log_{16} 12$

4. $\log 3 - \log 2$

5. $\log 14^6$

6. $\log_{20} 10^{16}$

7. $\log_3 16 + \log_2 4$

8. $\log 10 + \log 10$

9. $\log 125$

10. $\log_2 2^4$

Name________________________________ HSF-LE.A.4, HSF-BF.B.5

Simplifying and Solving Logarithms

Simplify each expression, then solve. Place the letter of the correct answer above the problem number below.

Example 1: $\log_3 x - \log_3 4 = \log_3 12$

$\log_3\left(\frac{x}{4}\right) = \log_3(12)$

therefore $\frac{x}{4} = 12$

$x = 48$

Example 2: $\log_5 7 + \frac{1}{2}\log_5 4 = \log_5 x$

$\log_5 7 + \log_5 4^{\frac{1}{2}} = \log_5 x$

$\log_5 7 + \log_5 2 = \log_5 x$

$\log_5 14 = \log_5 x$

$x = 14$

1. $\log_3 x - 2\log_3 2 = 3\log_3 3$
   M. 23
   N. 108
   O. $6\frac{3}{4}$

2. $\log_2 x = 9$
   A. 18
   E. 512
   I. 81

3. $\log_2 128 = x$
   C. 16
   D. 64
   E. 7

4. $\log_x 144 = {}^-2$
   N. 12
   O. 72
   P. $\frac{1}{12}$

5. $\log_2 x = \frac{1}{3}\log_2 27$
   N. 3
   O. 9
   P. 27

6. $\log_{16} 32 - \log_{16} 2 = x$
   W. 2
   X. 1
   Y. 16

7. $5\log 2 = \log x$
   E. 10
   I. 16
   O. 32

8. $\log_2 x - \log_2 5 = \log_2 10$
   R. 25
   S. 15
   T. 50

A logarithm is an __ __ __ __ __ __ __ __ .
2 6 4 7 5 3 1 8

Name________________________________

HSF-LE.A.4, HSF-BF.B.5

Using Logarithms to Solve $B^x = A$

Solve for x, rounding to the nearest tenth.

Example: $5^x = 30$

$\log 5^x = \log 30$

$x \cdot \log 5 = \log 30$

$x = \dfrac{\log 30}{\log 5} = 2.1113 = 2.1$

1. $9^x = 27$

2. $7^x = 343$

3. $10^x = 0$

4. $6^x = 127$

5. $12^x = 303$

6. $13^x = 2839$

7. $2^x = 90$

8. $4^x = 512$

9. $3^x = 5.2$

10. $11^x = 153$

Name________________________________ HSF-IF.A.1, HSF-IF.B.5

Domain and Range

Find the domain and range of each function.
Domain: Allowable values of *x* Range: *y* values

Example:

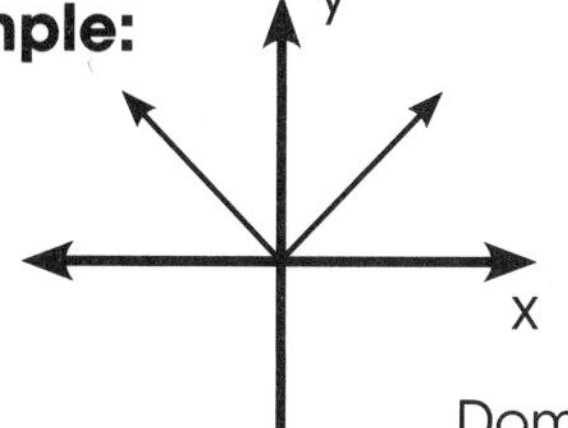

Domain: all Reals {ℝ}
Range: positive Reals (including 0) {ℝ: $y \geq 0$}

1.

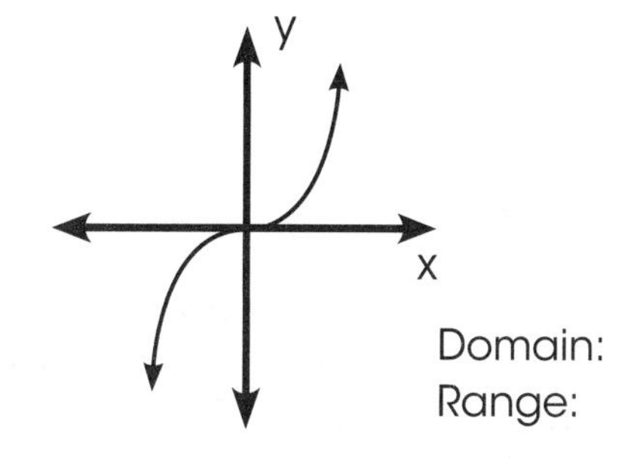

Domain:
Range:

2.

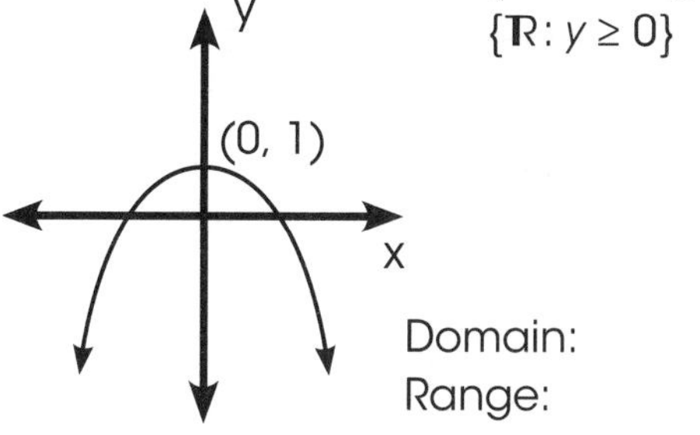

Domain:
Range:

3.

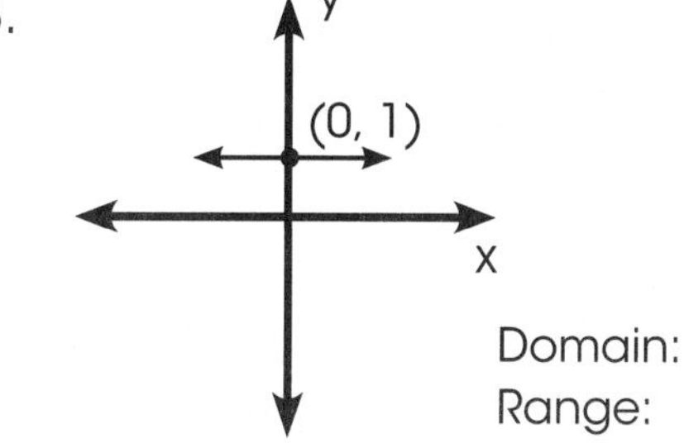

Domain:
Range:

4.

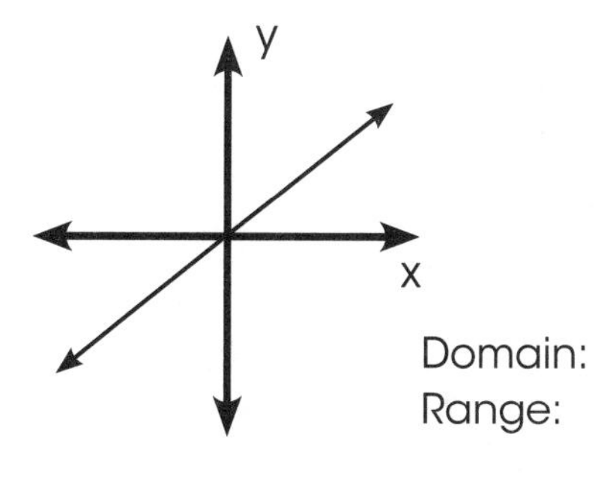

Domain:
Range:

5.

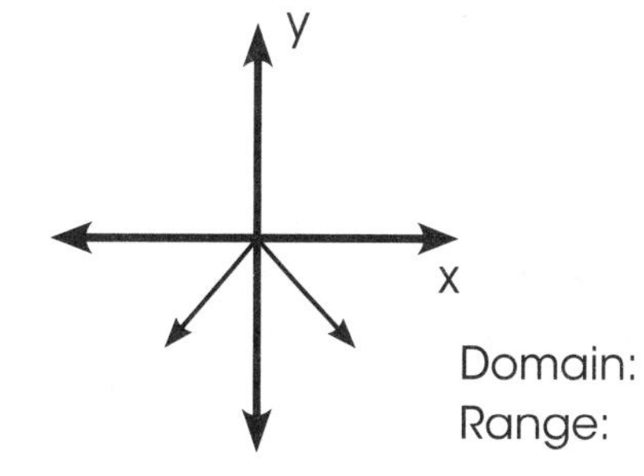

Domain:
Range:

6.

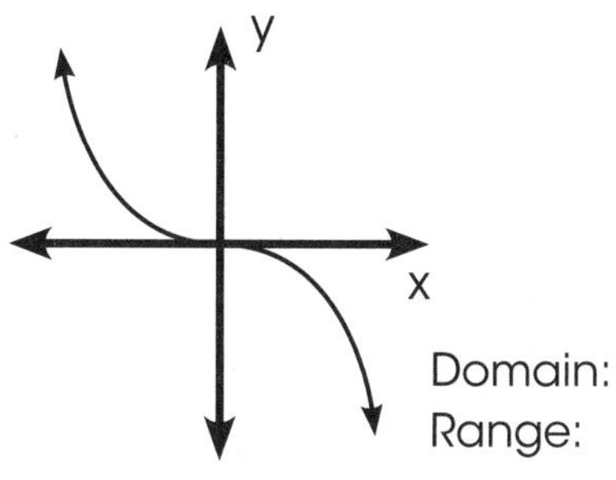

Domain:
Range:

7.

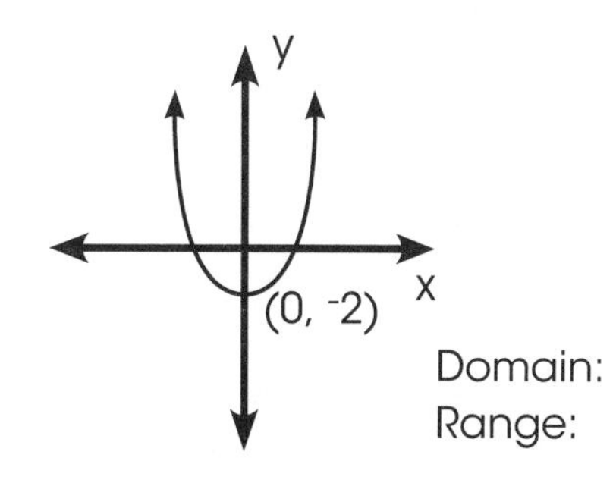

Domain:
Range:

8.

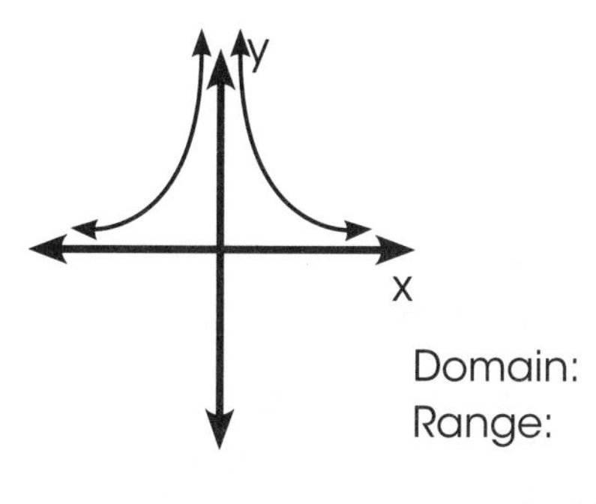

Domain:
Range:

9.

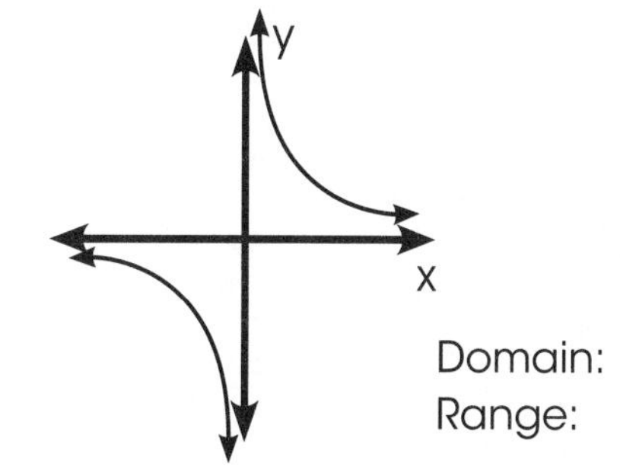

Domain:
Range:

Name____________________ HSF-IF.A.1, HSF-IF.B.5

More with Domain and Range

Find the domain and range from the graph.
Each box on the graph equals 1 unit.

Example:

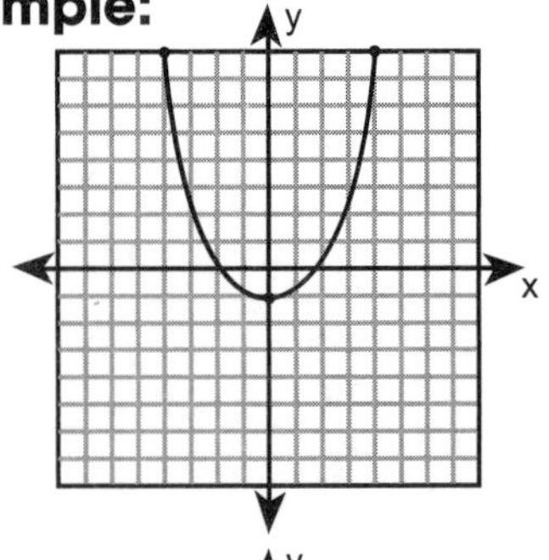

Domain: $^{-}4 \leq x \leq 4$
Range: $^{-}1 \leq y \leq 8$

1.

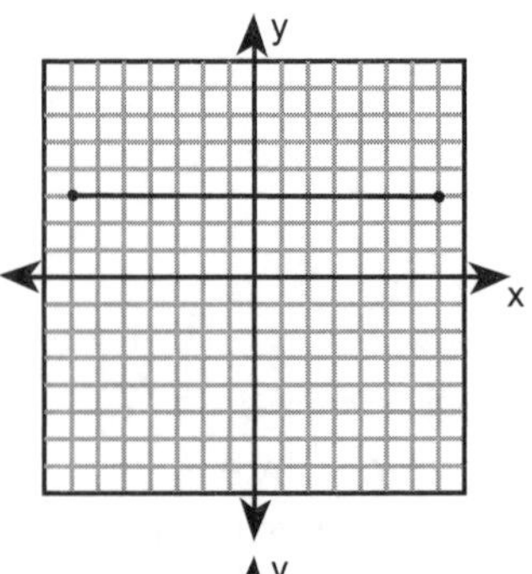

Domain:
Range:

2.

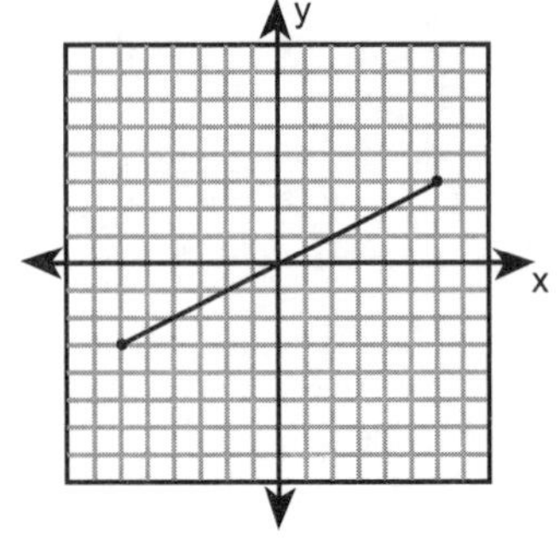

Domain:
Range:

3.

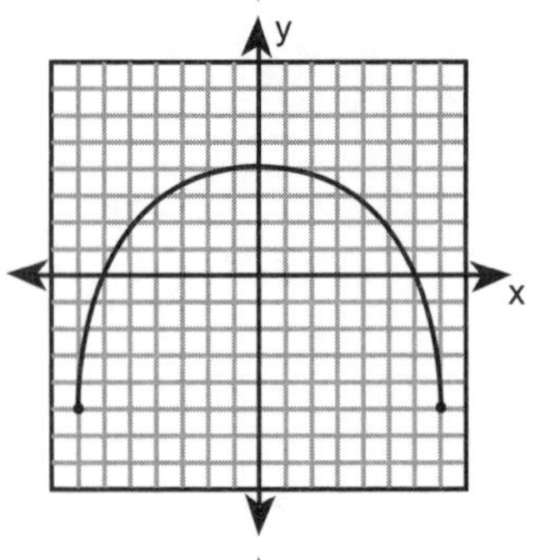

Domain:
Range:

4.

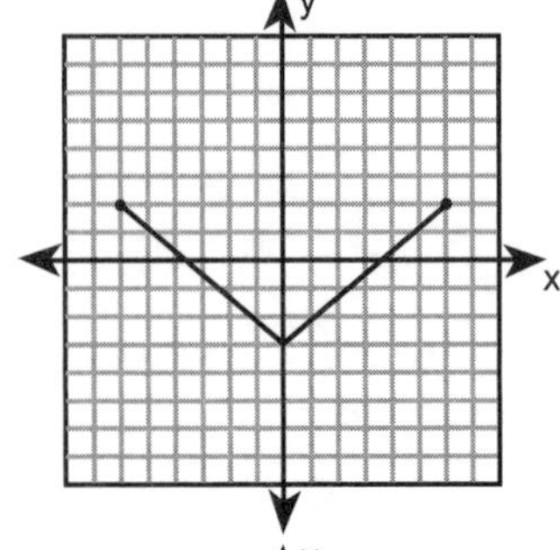

Domain:
Range:

5.

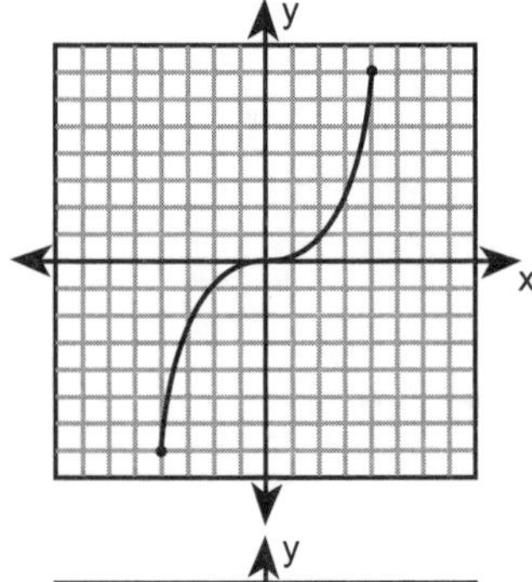

Domain:
Range:

6.

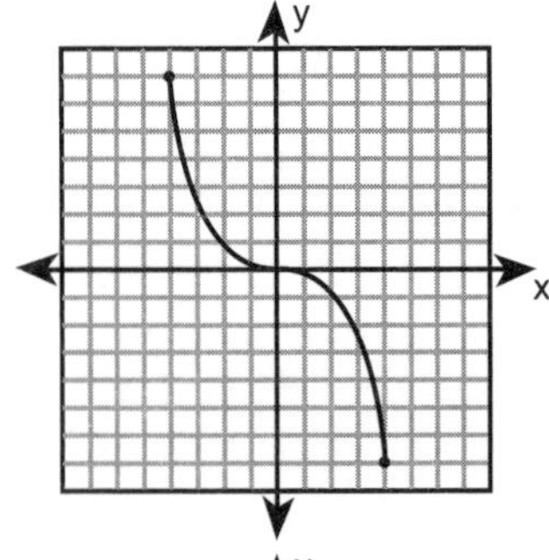

Domain:
Range:

7.

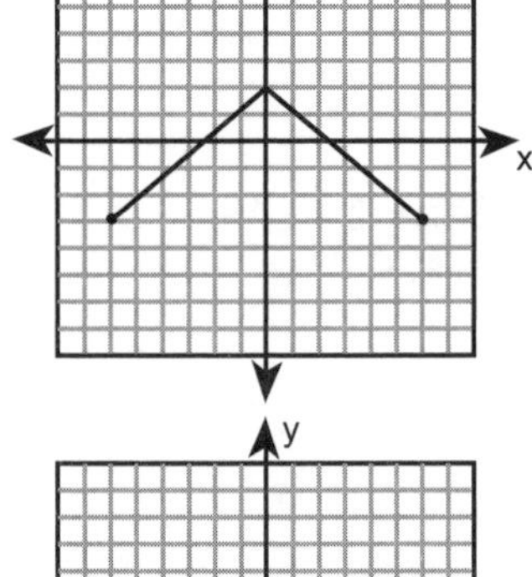

Domain:
Range:

8.

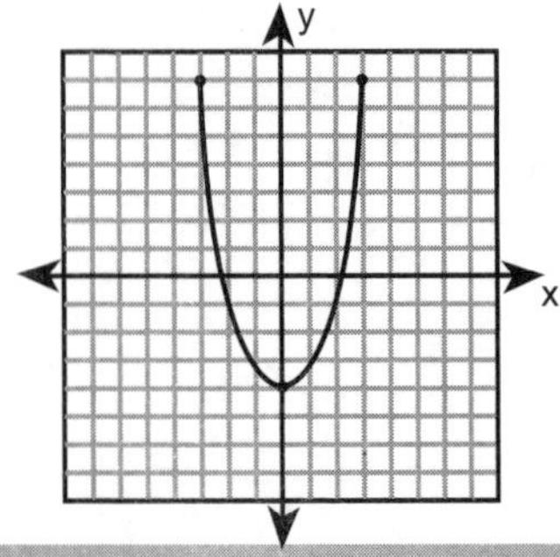

Domain:
Range:

9.

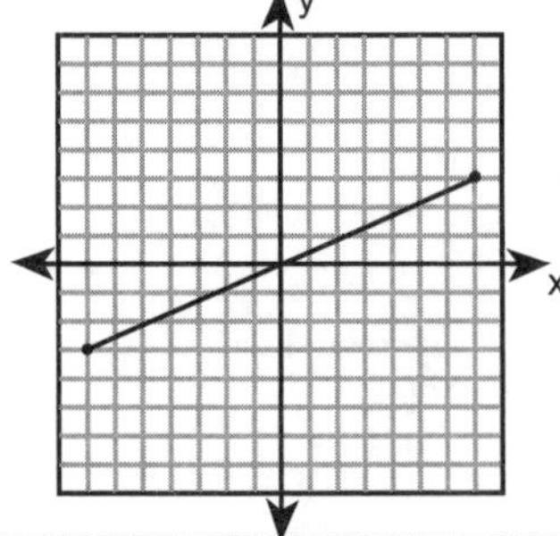

Domain:
Range:

Name______________________________

Just for Fun

Clark, Jones, Morgan, and Smith are four people whose occupations are salesperson, pharmacist, grocer, and police officer, though not necessarily in that order. Use the following statements to determine each person's occupation.

1. Clark and Jones are neighbors and take turns driving each other to work.
2. The grocer makes more money than Morgan.
3. Clark beats Smith regularly at bowling.
4. The salesperson always walks to work.
5. The police officer does not live near Clark.
6. The only time Morgan and the police officer ever met was when Morgan was stopped for speeding.
7. The grocer doesn't bowl.

	Salesperson	Pharmacist	Grocer	Police Officer
Clark				
Jones				
Morgan				
Smith				

Name____

HSG-SRT.C.6

Trigonometric Ratios

Use this helpful mnemonic to remember the following ratios:
Oscar **H**as **A** **H**eap **O**f **A**pples.

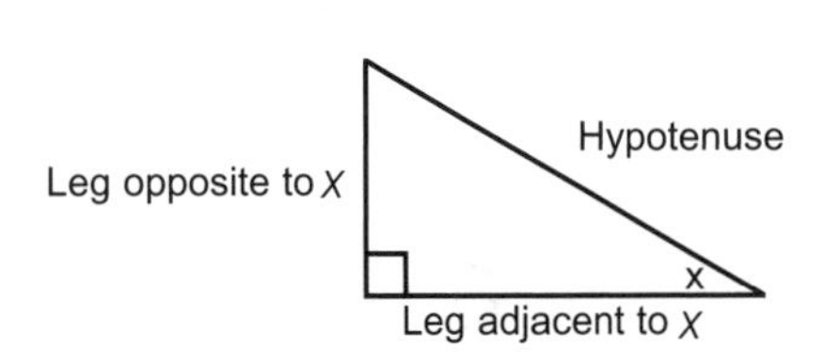

$$\text{Sine } x = \frac{\textbf{O}\text{pposite leg}}{\textbf{H}\text{ypotenuse}}$$

$$\text{Cosine } x = \frac{\textbf{A}\text{djacent leg}}{\textbf{H}\text{ypotenuse}}$$

$$\text{Tangent } x = \frac{\textbf{O}\text{pposite leg}}{\textbf{A}\text{djacent leg}}$$

Note: The trigonometric ratios hold only for right triangles.

Given a right triangle, find each trigonometric ratio. Leave your answer as a fraction. The first one has been started for you.

1. 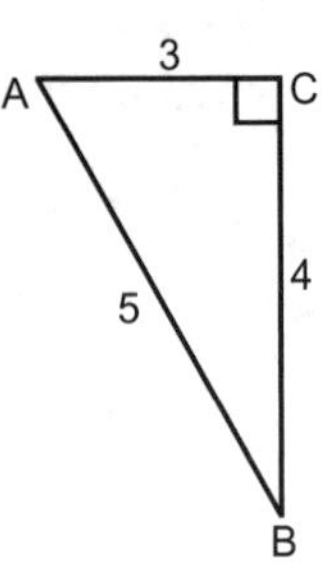

$\sin A = \frac{4}{5}$ $\sin B =$

$\cos A = \frac{3}{5}$ $\cos B =$

$\tan A = \frac{4}{3}$ $\tan B =$

2. 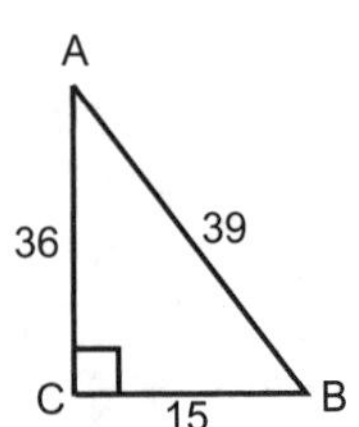

$\sin A =$ $\sin B =$

$\cos A =$ $\cos B =$

$\tan A =$ $\tan B =$

3. 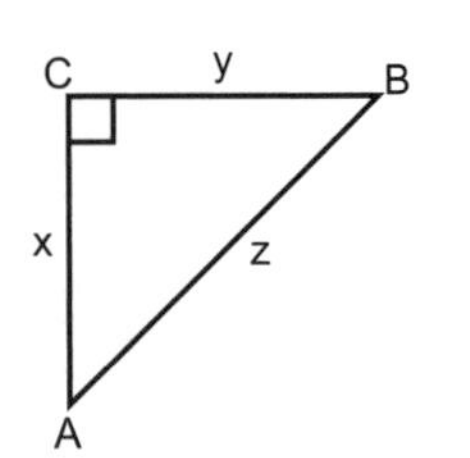

$\sin A =$ $\sin B =$

$\cos A =$ $\cos B =$

$\tan A =$ $\tan B =$

4. 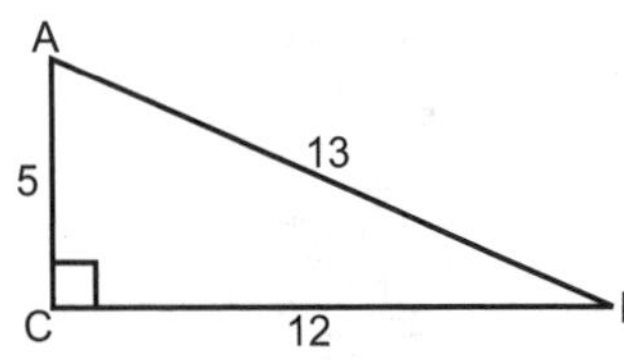

$\sin A =$ $\sin B =$

$\cos A =$ $\cos B =$

$\tan A =$ $\tan B =$

Name________________________________

HSG-SRT.C.6, HSG-SRT.C.7

Evaluating Trigonometric Functions

Evaluate each trigonometric function. Round to the nearest hundredth. You will need a scientific calculator.

Example: sin 60°
The calculator key sequence is [sin] [60] [=] 0.8660254° = 0.87°

Note: If you do not get the above answer, check the mode on your calculator. The mode should be in degrees. If you still do not get the correct answer, try [60] [sin] [=].

1. tan 45°
2. cos 10°
3. cos 220°
4. sin 80°
5. sin 23°
6. tan 135°

Find the angle with the given trigonometric ratio. Round your answer to the nearest degree.

Example: $\cos x = \left(\frac{6}{11}\right)$

calculator key sequence: [2nd] [cos] [6] [÷] [11] [=] 56.94426885° = 57°

Note: The mode on your calculator should still be in degrees. If you are not getting the correct answer, try [6] [÷] [11] [2nd] [cos] [=].

1. $\cos x = \left(\frac{7}{19}\right)$
2. $\tan x = \left(\frac{101}{90}\right)$
3. $\sin x = \left(\frac{20}{21}\right)$
4. $\cos x = \left(\frac{45}{76}\right)$
5. $\tan x = \left(\frac{15}{4}\right)$
6. $\sin x = \left(\frac{8}{99}\right)$

Name______________________________________ HSG-SRT.C.6, HSG-SRT.C.7

Applying Trigonometric Ratios

Using the trigonometric ratios, solve for the missing sides x and y of each right triangle. Round your answers to the nearest tenth.

Example:

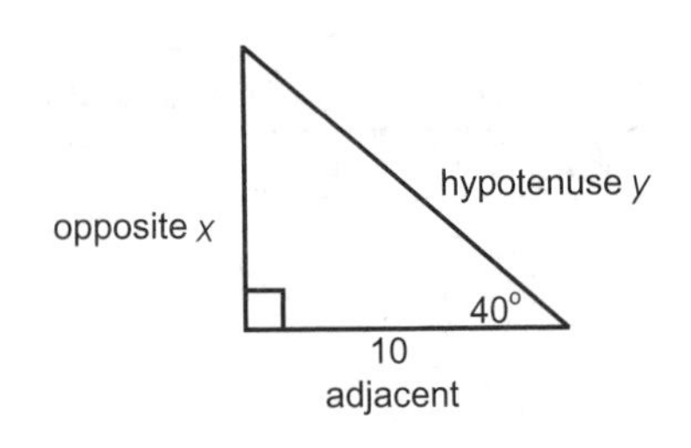

Ratios: $\sin 40 = \left(\frac{x}{y}\right)$ ← **Note:** having two variables in one ratio means it cannot be solved.

$\cos 40 = \left(\frac{10}{y}\right)$

$\tan 40 = \left(\frac{x}{10}\right)$

Use the tan and the cos to solve for x and y.

$\cos 40° = \frac{10}{y}$

$y \cos 40° = 10$

$y = \frac{10}{\cos 40°} = 13.05407$

$y = 13.1$

$\tan 40° = \frac{x}{10}$

$x = 10 \tan 40° = 8.39099$

$x = 8.4$

Note: Since these are right triangles, you can check your answer using the Pythagorean theorem. The answers will not be exact due to rounding.

1. 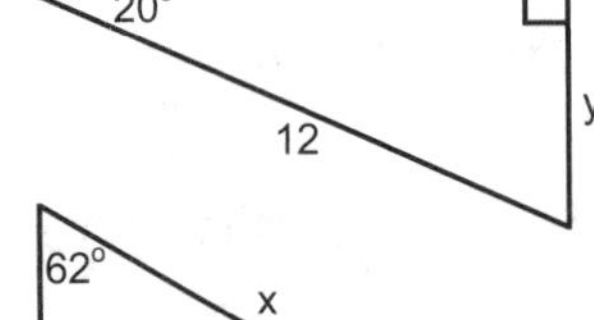

2. 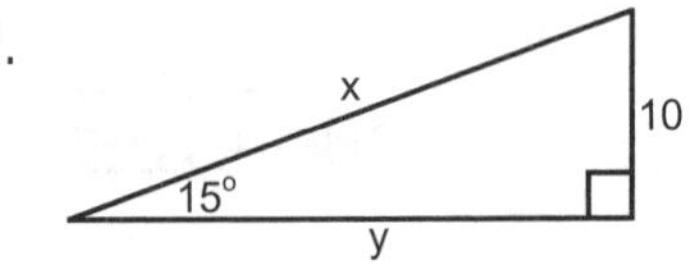

3. 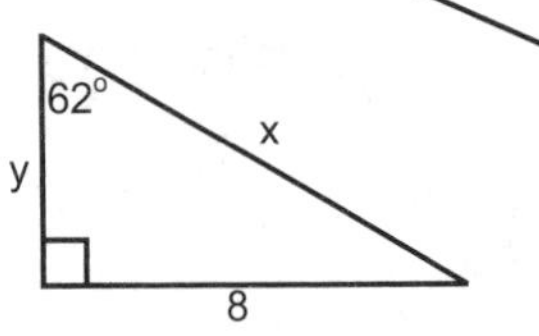

4. 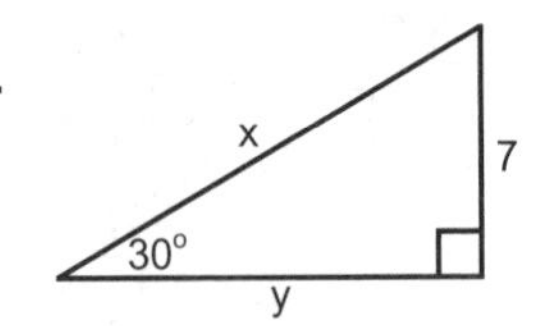

5. 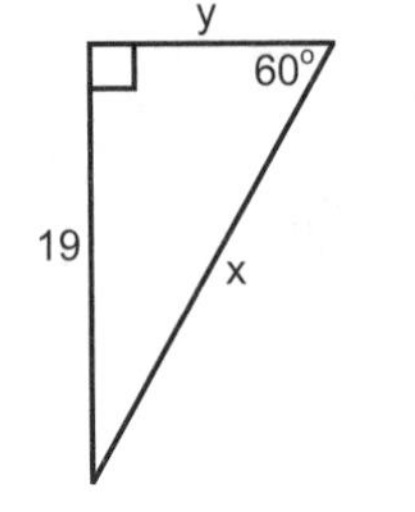

6. 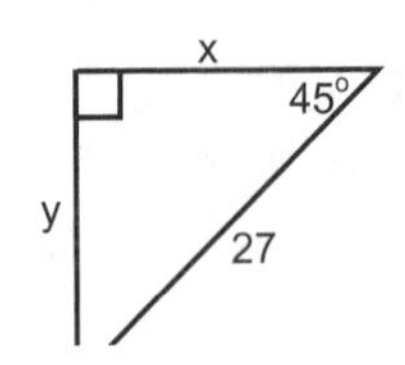

7. 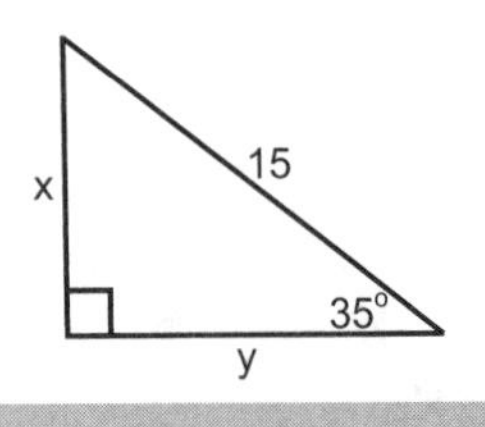

8. 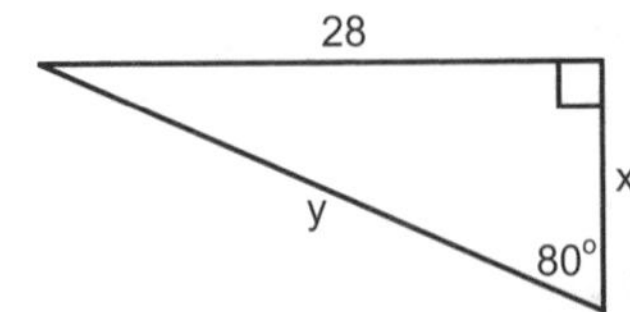

Name________________________________

HSG-SRT.C.6, HSG-SRT.C.7

Using Trigonometric Ratios to Find Angles

Using the trigonometric ratios, solve for the missing angles *x* and *y* of each right triangle. Round your answers to the nearest tenth.

Example:

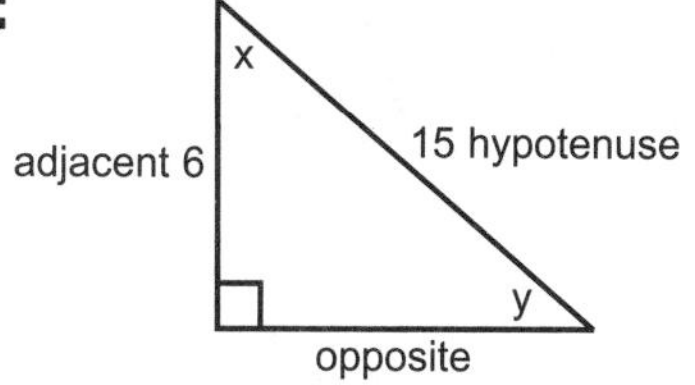

x
opposite 6
15 hypotenuse
y
adjacent

Solving for angle x:

$\cos x = \left(\frac{6}{15}\right)$

$x = \cos^{-1}\left(\frac{6}{15}\right)$

$x = 66.421 = 66°$

Solving for angle y:

$\sin y = \left(\frac{6}{15}\right)$

$y = \sin^{-1}\left(\frac{6}{15}\right)$

$y = 23.578 = 24°$

Check: 66° + 24° = 90°

Check: The three angles of a triangle always add up to 180 degrees. Since these are right triangles, one angle must equal 90 degrees. Therefore, the other two must add up to 90 degrees. Remember, rounding may cause the answers to be slightly off.

1. 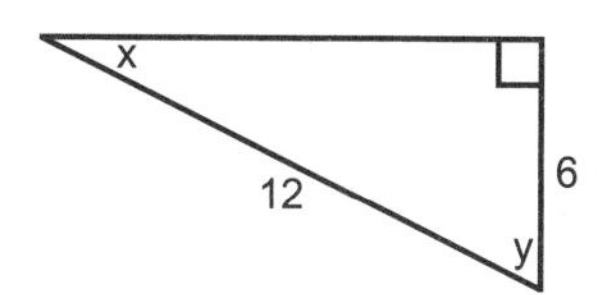

2. 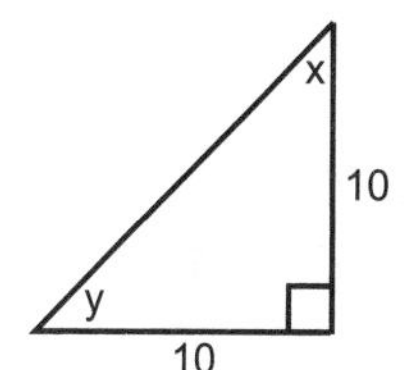

3. 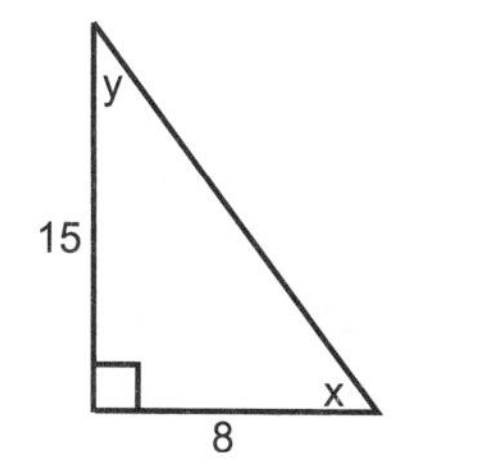

4. 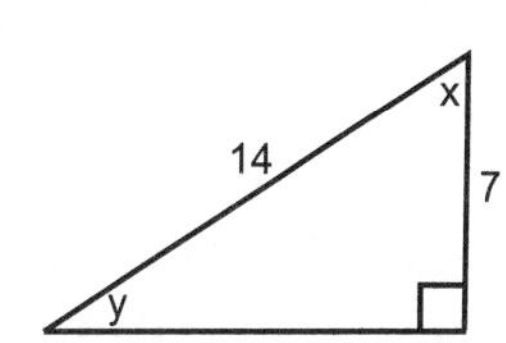

5. 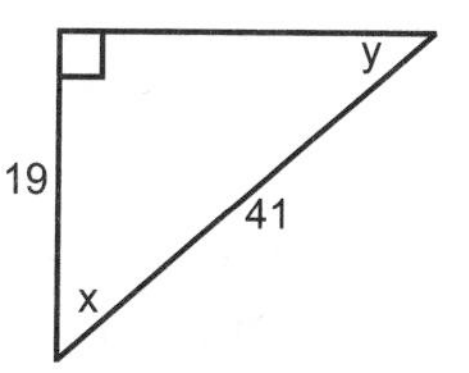

6. 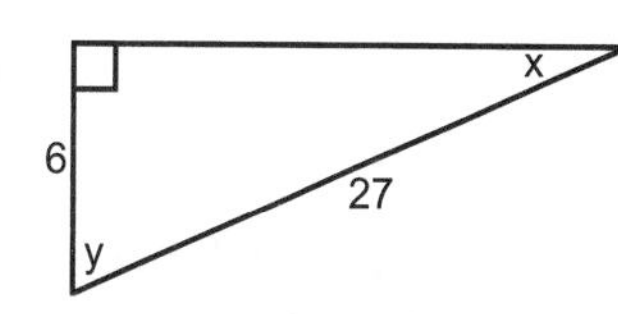

7. 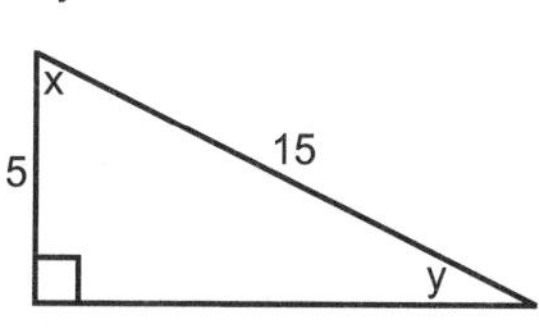

8. 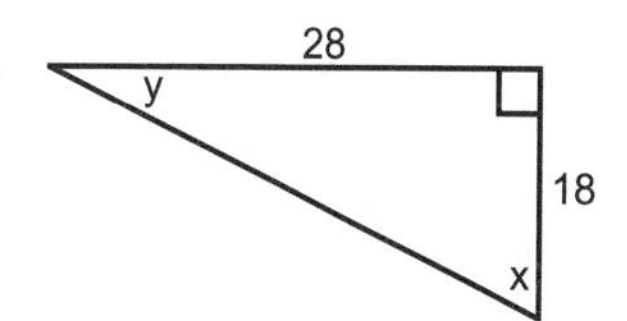

Name____

HSG-SRT.C.6, HSG-SRT.C.7

Problem Solving With Trigonometric Ratios

Draw a picture and solve the story problem using trigonometric ratios.

Example: An eagle spotted a mouse 20 feet below at an angle of 42 degrees with the horizon. If the eagle flies along its line of sight, how far will it have to fly to reach its prey?

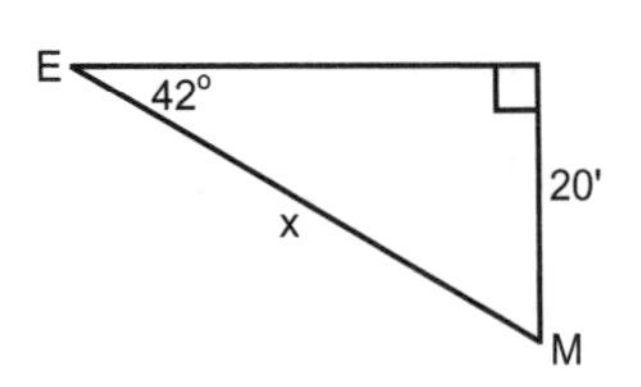

$$\sin 42° = \frac{20}{x}$$

$$x = \frac{20}{\sin 42} = 29.889 = 29.9 \text{ feet}$$

1. A 20-foot ladder is leaning against a wall. The base of the ladder is 3 feet from the wall. What angle does the ladder make with the ground?

2. How tall is a bridge if a 6-foot-tall person standing 100 feet away can see the top of the bridge at an angle of 30 degrees to the horizon?

3. An air force pilot must descend 1500 feet over a distance of 9000 feet to land smoothly on an aircraft carrier. What is the plane's angle of descent?

4. In a movie theater 150 feet long, the floor is sloped so there is a difference of 30 feet between the front and back of the theater. What is the angle of depression?

5. A bow hunter is perched in a tree 15 feet off the ground. If he sees his prey on the ground at an angle of 30 degrees, how far will the arrow have to travel to hit his target?

Name________________________________ HSF-TF.A.1, HSF-TF.A.2, HSF-TF.A.3

Creating the Unit Circle

One of the most useful tools in trigonometry is the unit circle. The unit circle is a circle with a radius of one unit placed on an *x*–*y* plane. The angles are measured from the positive *x*-axis counterclockwise. The *x*-axis corresponds to the cosine function and the *y*-axis corresponds to the sine function. To create the unit circle, one uses the special right triangles: 30° 60° 90° and 45° 45° 90°. Study the following diagrams to learn how to create the unit circle instead of merely memorizing it!

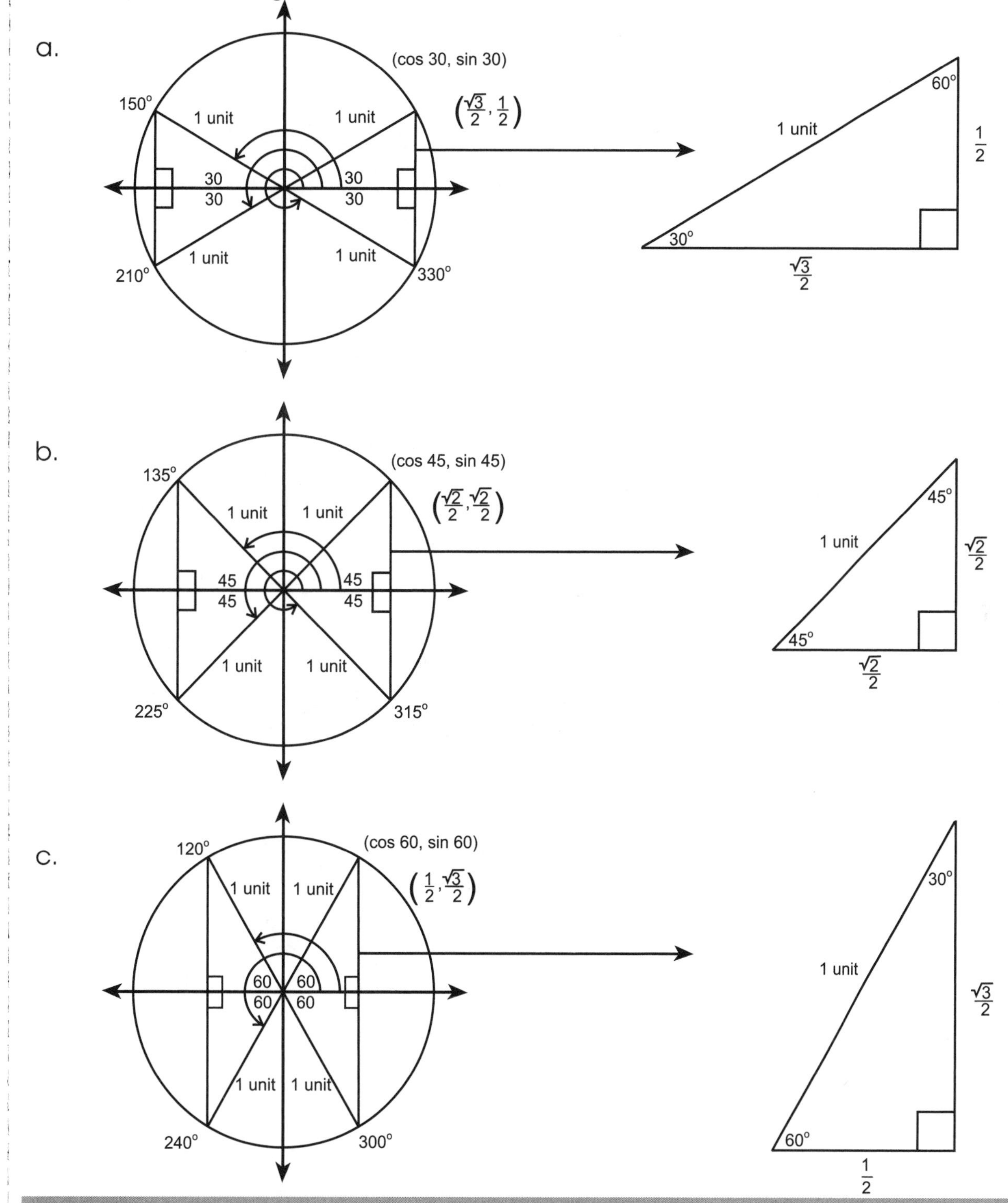

Name________________________ HSF-TF.A.1, HSF-TF.A.2, HSF-TF.A.3

More Creating the Unit Circle

Note: Three congruent right triangles are created in each different quadrant. Therefore, the length of the sides will be the same in each triangle. The only difference will be the sign (positive or negative) of the numbers.

Complete the unit circle below, being watchful of the signs of the numbers in each quadrant.

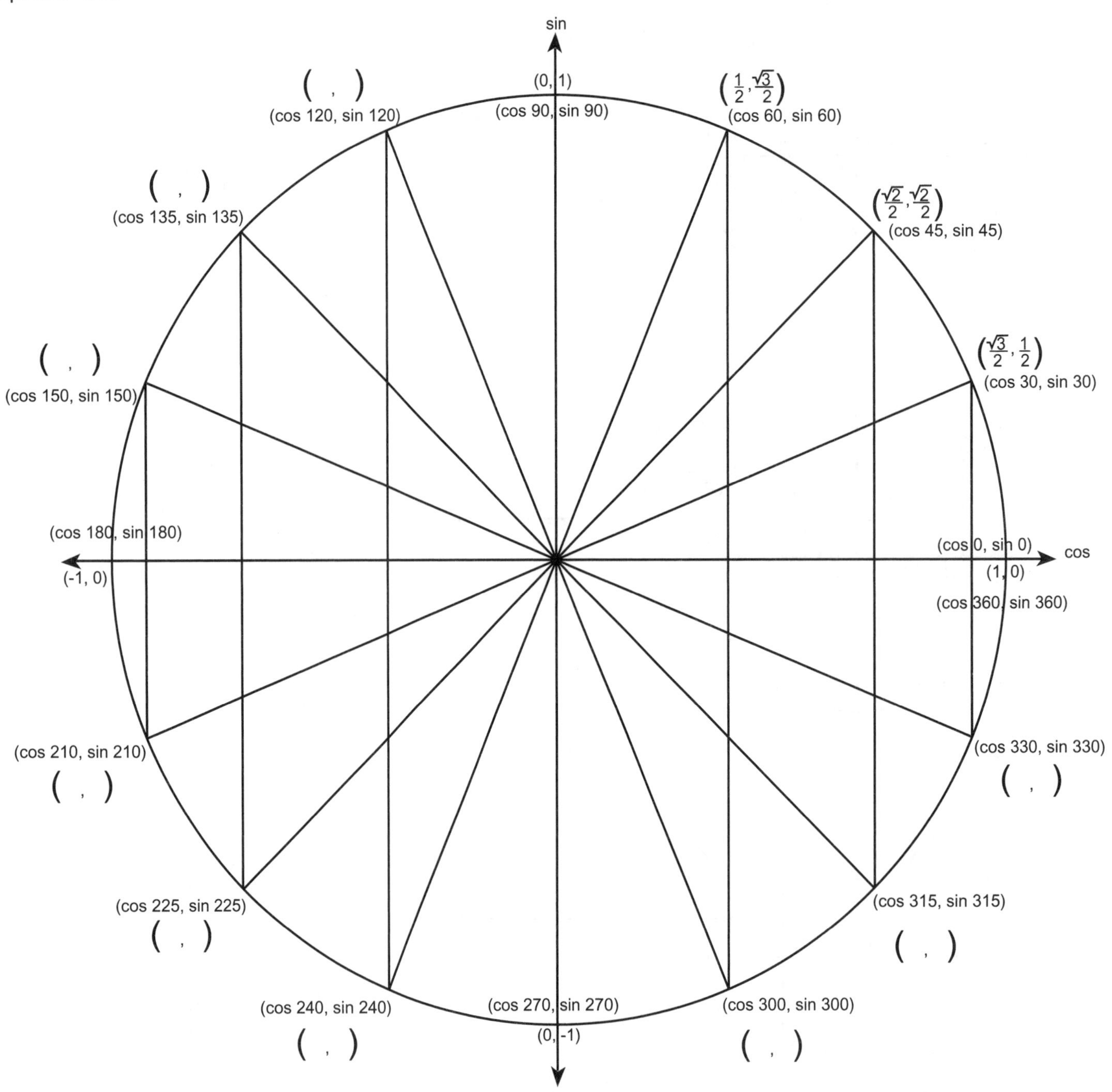

Name____________________

HSF-TF.A.1, HSF-TF.A.2, HSF-TF.A.3

Unit Circle With Negative Angle Measures

Recall that the angles of the unit circle are measured from the positive *x*-axis counterclockwise. The angles can be measured from the positive *x*-axis going in the clockwise direction, but in such a case the sign of the angle measured is negative.

Complete the unit circle using negative angle measures.

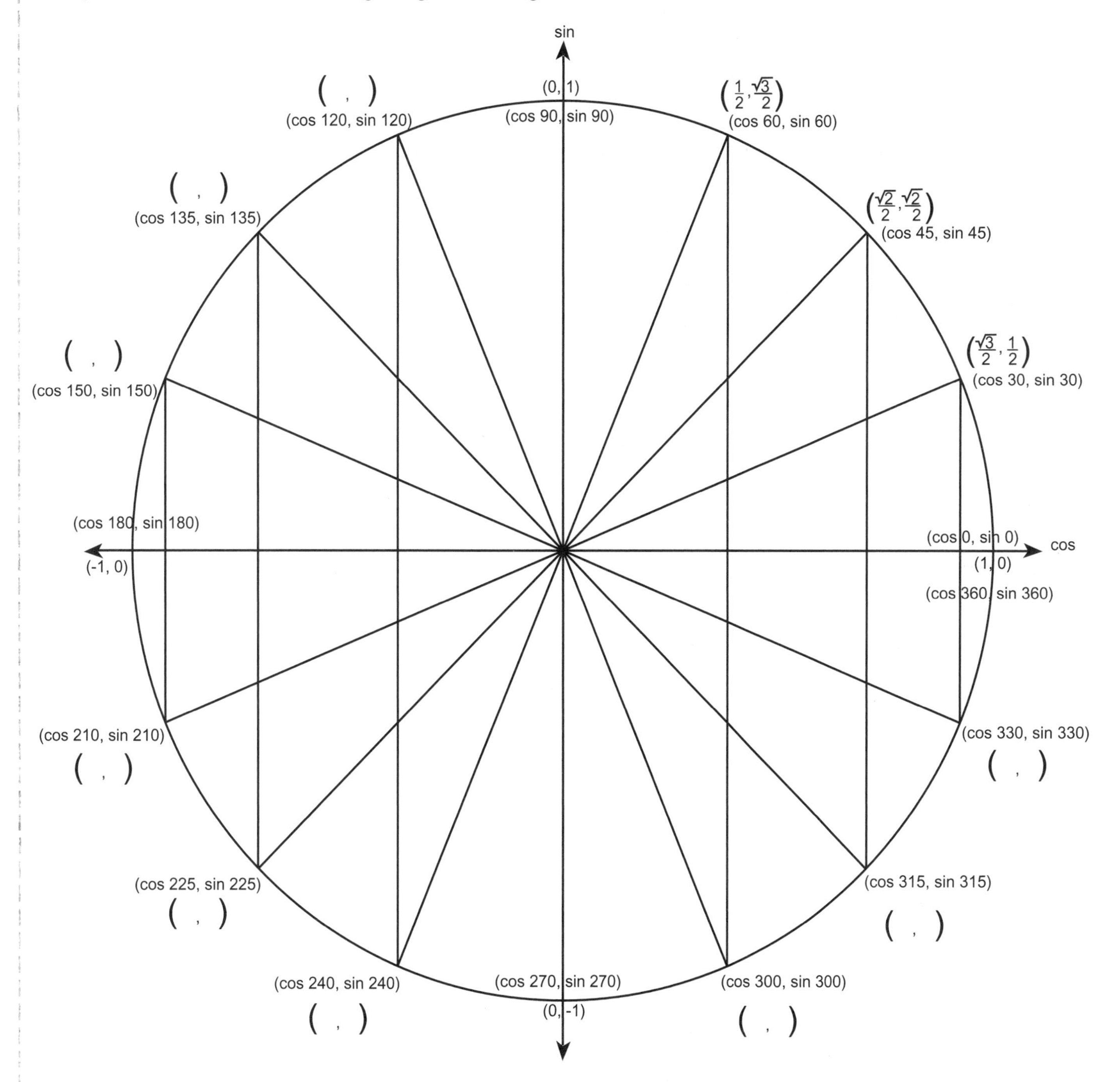

What did you notice about the positive and negative angles measured at each point?

Name________________________________

HSF-TF.A.3, HSF-TF.C.8

Angles Greater Than 360 Degrees

Without a calculator, evaluate the following trigonometric functions.
Hint: Rewrite each statement using an angle ≤ 360°.

Example: sin 450°

360 degrees represents one complete revolution about the unit circle.
So, 450 degrees is one complete revolution plus 90 degrees more in the counterclockwise direction.

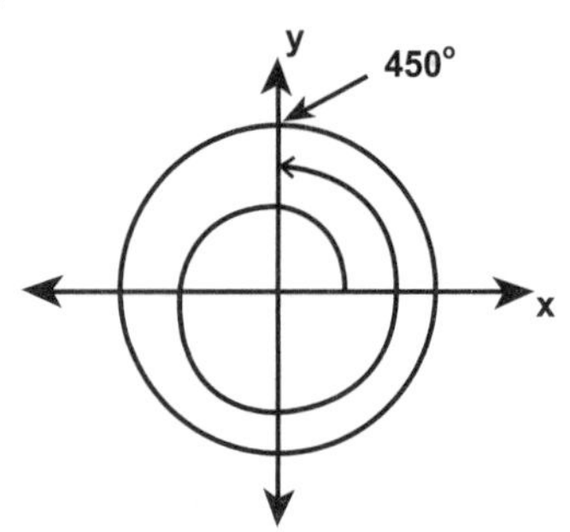

450 – 360 = 90
The problem reduces to sin 90 = 1.

Note: Refer to the unit circle diagram while you complete this assignment.

1. sin 630°
2. sin $^-$750°
3. cos $^-$480°
4. sin $^-$420°
5. sin 510°
6. cos 1020°
7. cos $^-$540°
8. cos $^-$675°
9. sin $^-$540°
10. cos $^-$930°
11. sin 405°
12. sin $^-$600°
13. sin 3600°
14. cos $^-$1830°

Each sine problem has the same value as one of the cosine problems. List the pairs.

1 & ____; 2 & ____; 4 & ____; 5 & ____; 9 & ____; 11 & ____; 12 & ____

In each pair, what is the relationship of the reference angles? ______________________________

____

Name________________________________

HSF-TF.A.3, HSF-TF.C.8

Converting Angle Measurements

On the unit circle, the angles are given in degrees. Another way to measure an angle is with radians. π Radians = 180°. The two unit conversions are as follows:

radians to degrees	degrees to radians
$\frac{180°}{\pi}$	$\frac{\pi}{180°}$

Convert the angle from degrees to radians, leaving your answer as a fraction.

Example: 90° Simply multiply. $\frac{90°}{1} \times \frac{\pi}{180°} = \frac{90°\pi}{180°} = \frac{1}{2}\pi$

1. 310°
2. 150°
3. 30°
4. 420°
5. 120°
6. 350°

Convert from radians to degrees.

Example: $\frac{3}{2}\pi$ Simply multiply. $\frac{3}{2}\pi \times \frac{180°}{\pi} = \frac{540°}{2} = 270°$

1. $\frac{5}{4}\pi$
2. 4π
3. $\frac{7}{6}\pi$
4. $\frac{1}{6}\pi$
5. $\frac{7}{4}\pi$
6. $\frac{9}{2}\pi$

Name____________________________________ HSF-TF.A.3, HSF-TF.C.8

Unit Circle With Radian Angle Measures

Complete the unit circle using radian angle measures. Use as a reference when completed.

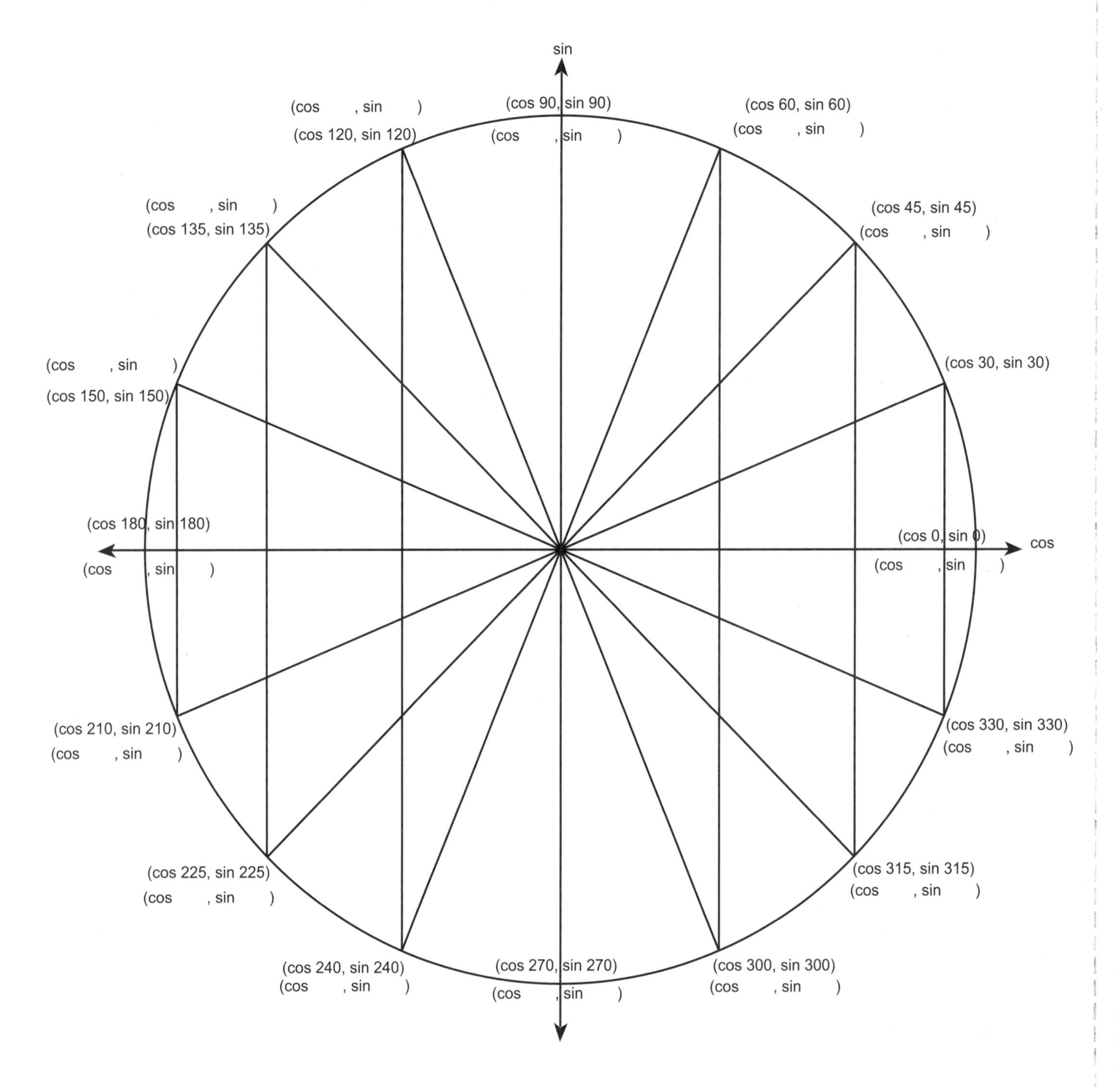

Name________________________________

HSF-TF.C.9

Manipulating Properties of Sine and Cosine

General Properties | **Examples:**

$\sin x = \cos(90° - x)$ ⟶ $\sin 30° = \cos(90° - 30°) = \cos 60°$

$\cos x = \sin(90° - x)$ ⟶ $\cos 75° = \sin(90° - 75°) = \sin 15°$

$\sin x = \sin(180° - x)$ ⟶ $\sin 45° = \sin(180° - 45°) = \sin 135°$

$\cos x = \cos(360° - x)$ ⟶ $\cos 120° = \cos(360° - 120°) = \cos 240°$

Given the sin x, name two angles such that $\sin x = \cos y$.

1. $\sin 60°$
2. $\sin 15°$
3. $\sin 180°$
4. $\sin {}^{-}45°$
5. $\sin {}^{-}120°$

Given cos x, name two angles such that $\cos x = \sin y$.

1. $\cos 30°$
2. $\cos {}^{-}60°$
3. $\cos 90°$
4. $\cos {}^{-}225°$
5. $\cos 50°$

Name____ HSF-TF.B.6, HSF-TF.B.7

Graphing the Sine and Cosine Functions

Graph, using a table of values over the specified domain. Add units to both axes.

Example: $y = \sin x$.
where $0° \leq x \leq 360°$

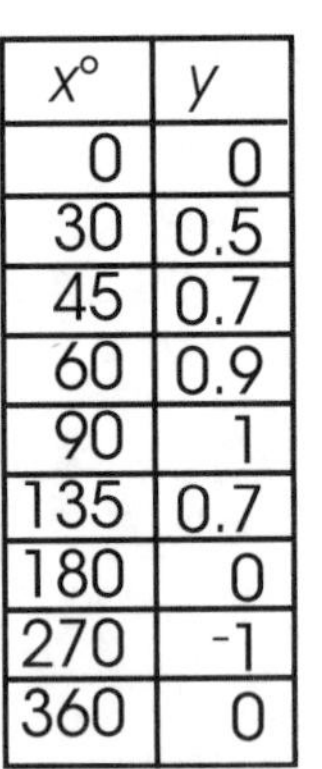

x°	y
0	0
30	0.5
45	0.7
60	0.9
90	1
135	0.7
180	0
270	⁻1
360	0

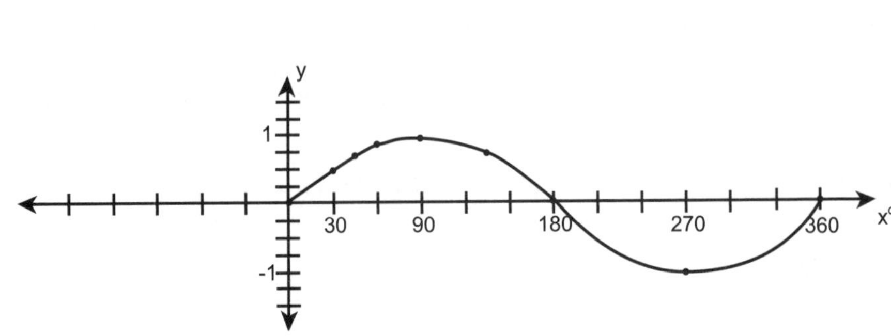

1. $y = \cos x$
where $0° \leq x \leq 360°$

x°	y
0	
30	
45	
60	
90	
135	
180	
270	
360	

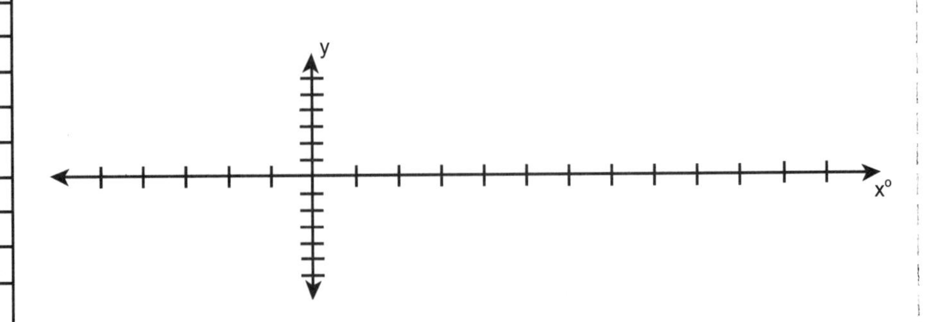

2. $y = \sin x$
where $90° \leq x \leq 450°$

x°	y

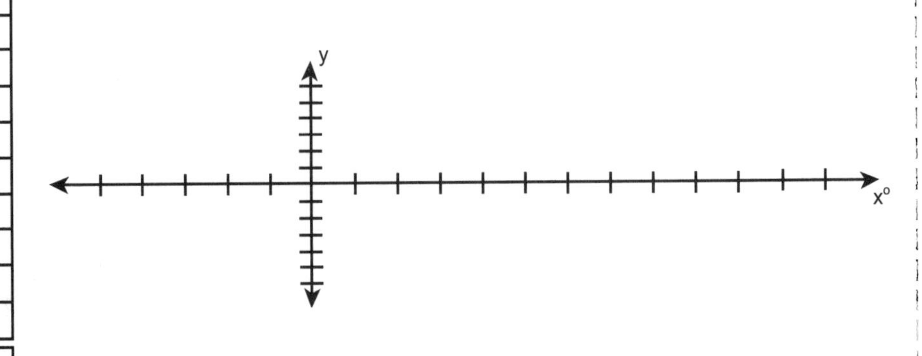

3. $y = \sin x$
where $^-90° \leq x \leq 270°$

x°	y

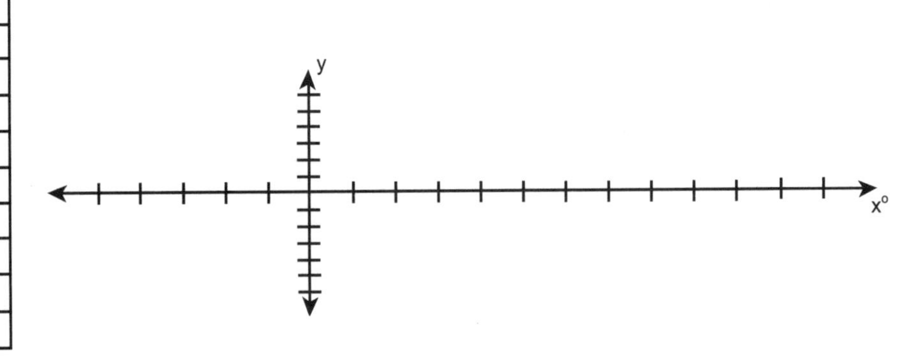

What do you notice about graphs #1 and #2?

Name____

HSF-TF.B.6, HSF-TF.B.7

More Graphing Sine and Cosine Functions

4. $y = \sin x$
where $^-360° \leq x \leq 360°$

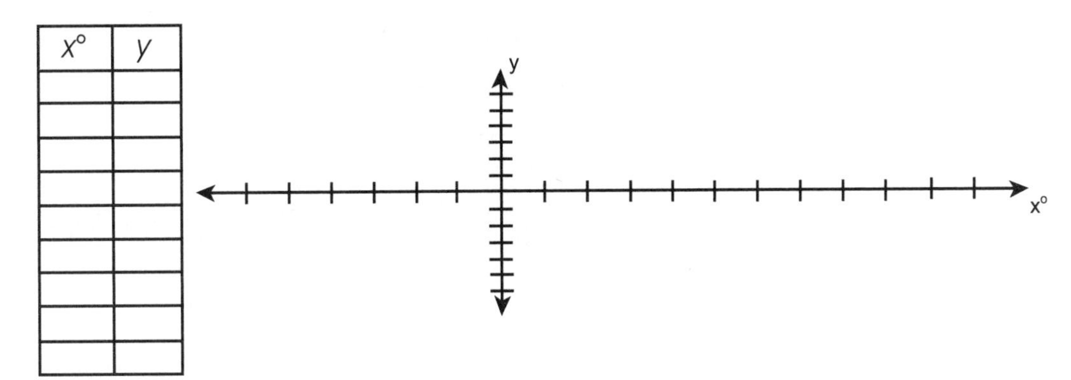

5. $y = \cos x$
where $^-360° \leq x \leq 360°$

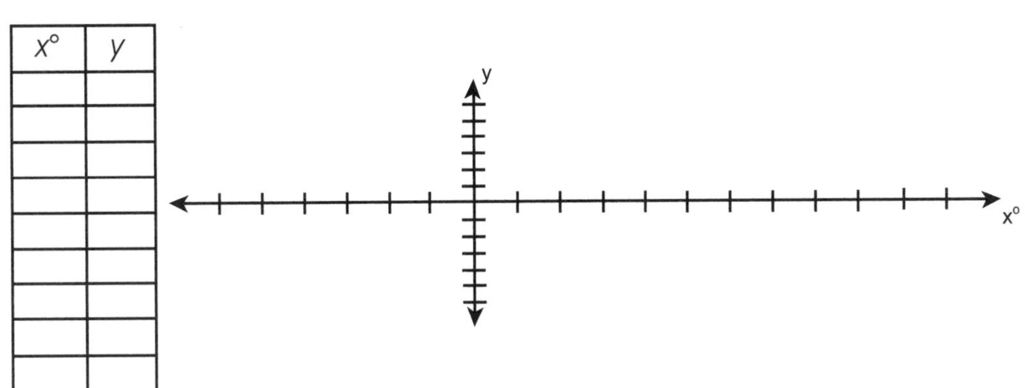

6. $y = \sin x$
where $0° \leq x \leq 720°$

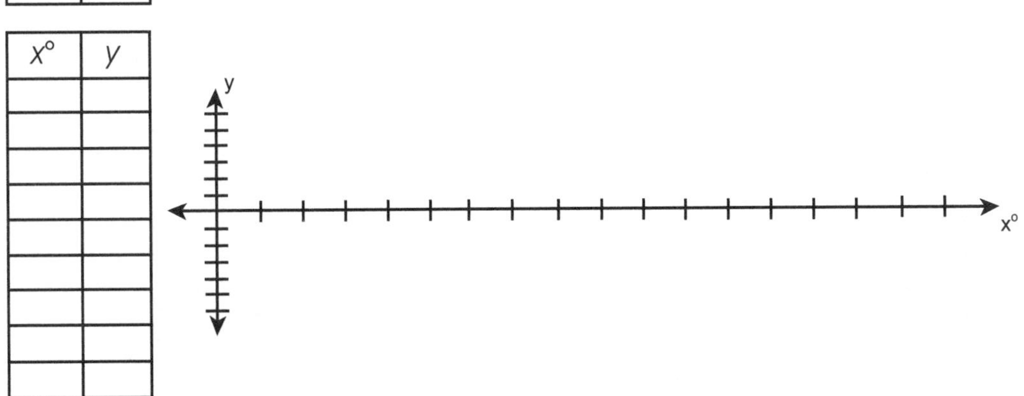

7. $y = \cos x$
where $0° \leq x \leq 720°$

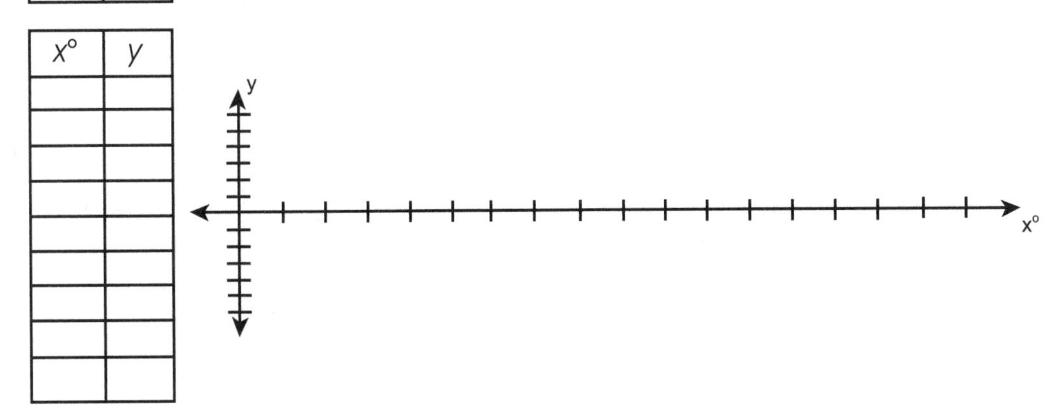

Name________________________________

Graphing y = a sin x or y = a cos x

Graph each function. Add units to both axes.

Example: $y = {}^{-}2 \cos x$

x°	y
0	$^{-}2$
90	0
180	2
270	0
360	$^{-}2$

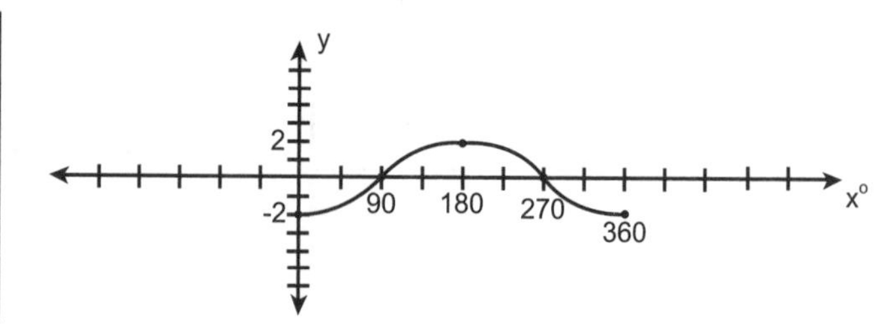

1. $y = 4 \cos x$

x°	y
0	
90	
180	
270	
360	

2. $y = \frac{3}{2} \sin x$

x°	y
0	
90	
180	
270	
360	

3. $y = \frac{1}{2} \cos x$

x8	y
0	
90	
180	
270	
360	

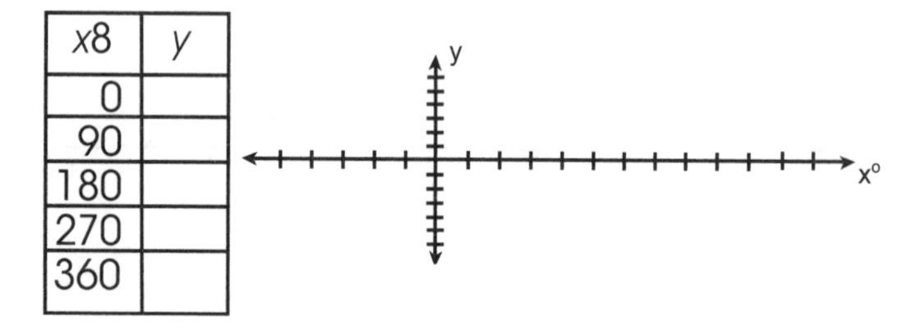

4. $y = {}^{-}3 \sin x$

x°	y
0	
90	
180	
270	
360	

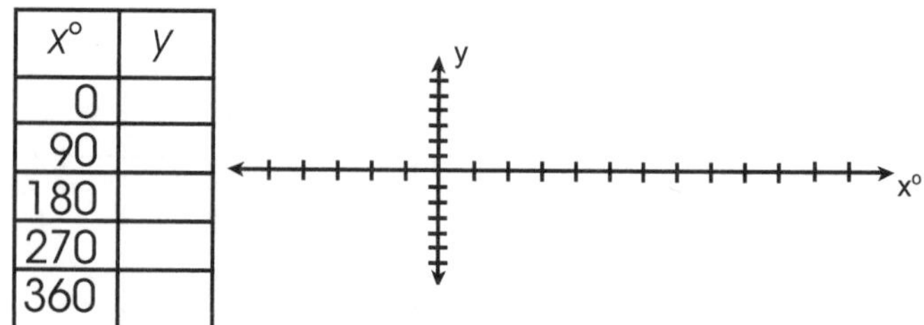

5. $y = 4 \sin x$

x°	y
0	
90	
180	
270	
360	

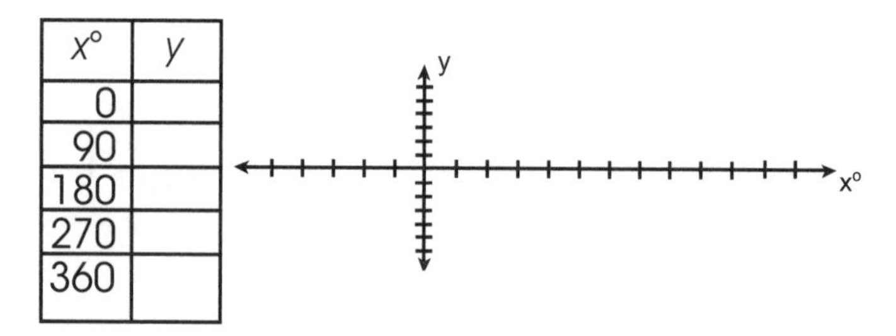

6. $y = -\frac{1}{3} \sin x$

x°	y
0	
90	
180	
270	
360	

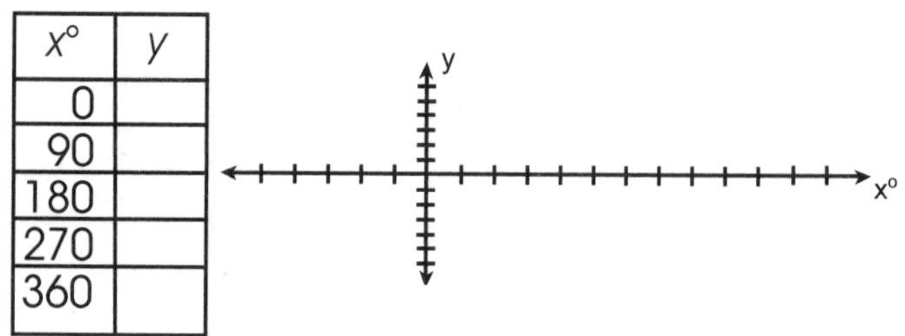

What does *a* do to the sin and cos function?
What does -*a* do to the sin and cos function?

Name________________________________ HSF-TF.B.6, HSF-TF.B.7

Graphing $y = c + a \sin x$ or $y = c + a \cos x$

Graph each function. Add units to both axes.

Example: $y = 1 + \sin x$

$x°$	y
0	$^-2$
90	0
180	2
270	0
360	$^-2$

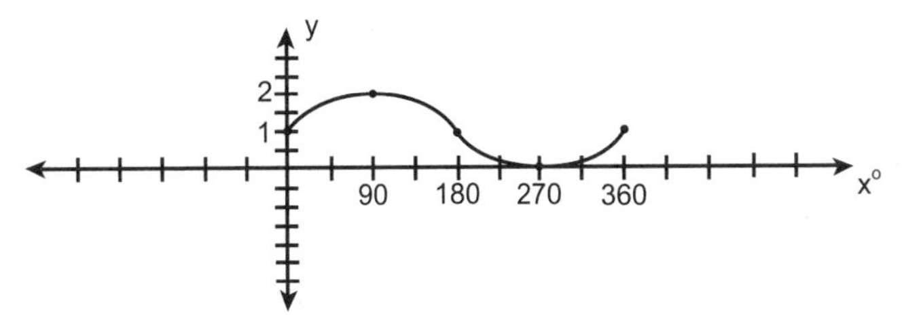

1. $y = 3 + \cos x$

$x°$	y
0	
90	
180	
270	
360	

2. $y = 2 + \sin x$

$x°$	y
0	
90	
180	
270	
360	

3. $y = \frac{1}{2} + \cos x$

$x°$	y
0	
90	
180	
270	
360	

4. $y = {}^-3 + \sin x$

$x°$	y
0	
90	
180	
270	
360	

5. $y = {}^-1 + \cos x$

$x°$	y
0	
90	
180	
270	
360	

6. $y = -\frac{1}{3} + \sin x$

$x°$	y
0	
90	
180	
270	
360	

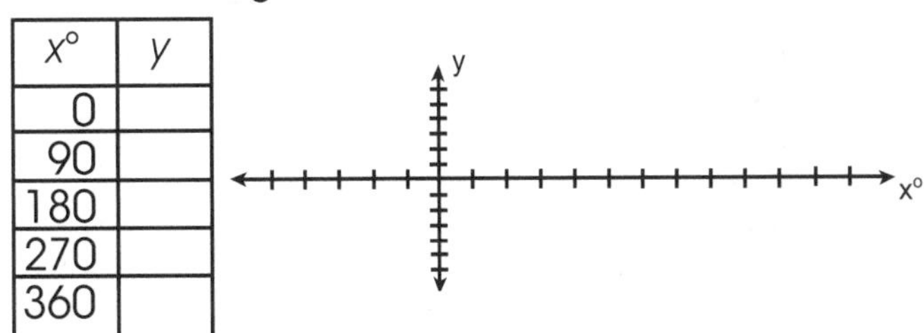

What does a positive c value do?
What does a negative c value do?

Name________________________________

Graphing *y* = sin *bx* or *y* = cos *bx*

The normal period, one complete wave, of a sine or cosine function is 360 degrees. However, the period will change due to the constant b in the general equation, $y = \sin bx$, $y = \cos bx$. In general, the period will be $\frac{360}{b}$. For example, the equation $y = \sin 2x$ has a period of $\frac{360}{2}$ or 180 degrees, that is, one complete sine wave in 180 degrees or two complete sine waves in 360 degrees. Refer to the table of values and the graph below. Notice that the table of values has three columns instead of the usual two. The middle column is always bx, and the usual values of 0, 90, 180, 270, and 360 are placed here. The first column contains the values in the middle column divided by b (or in this case, two). The third column contains the value of the function y. **Note:** To graph the function, use only the first and last columns of the table of values.

Complete the table of values and graph the function. Write the units on both axes.

Example: $y = \sin 2x$

$x°$	$2x°$	y
0	0	0
45	90	1
90	180	0
135	270	-1
180	360	0

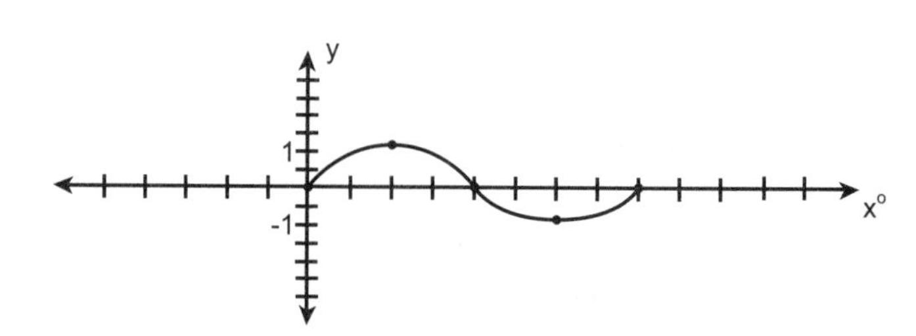

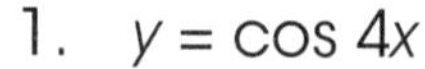

1. $y = \cos 4x$

$x°$		y
	0	
	90	
	180	
	270	
	360	

2. $y = \sin 3x$

$x°$		y
	0	
	90	
	180	
	270	
	360	

3. $y = \sin \frac{1}{2} x$

$x°$		y
	0	
	90	
	180	
	270	
	360	

4. $y = \cos \frac{1}{3} x$

$x°$		y
	0	
	90	
	180	
	270	
	360	

5. $y = \sin \frac{1}{4} x$

$x°$		y
	0	
	90	
	180	
	270	
	360	

6. $y = \cos 2x$

$x°$		y
	0	
	90	
	180	
	270	
	360	

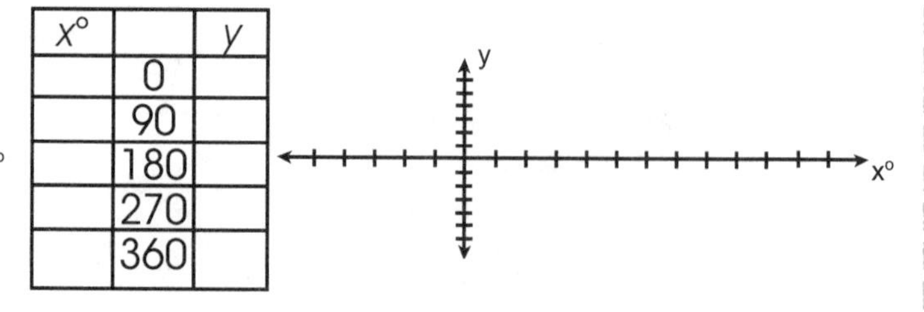

Name____________________________________ HSF-TF.B.6, HSF-TF.B.7

Graphing $y = \sin(x - d)$ or $y = \cos(x - d)$

In the general equation $y = \sin(x - d)$ or $y = \cos(x - d)$, $(x - d)$ is known as the argument of the function. $(x - d)$ corresponds to a phase shift or horizontal shift of d degrees. To graph functions of this type, a table of values with three three columns is used. The first column is for the variable. The middle column is for the argument $(x - d)$, and the last column is for the value of the function y. **Note:** Use only the first and last columns to graph the function. The numbers in the middle column insure that the graph will have one complete sine or cosine wave.

Complete the table of values and graph the function. Write the units on both axes.

Example: $y = \sin(x - 90°)$

x°	(x-90)	y
0	0	0
45	90	1
90	180	0
135	270	-1
180	36	0

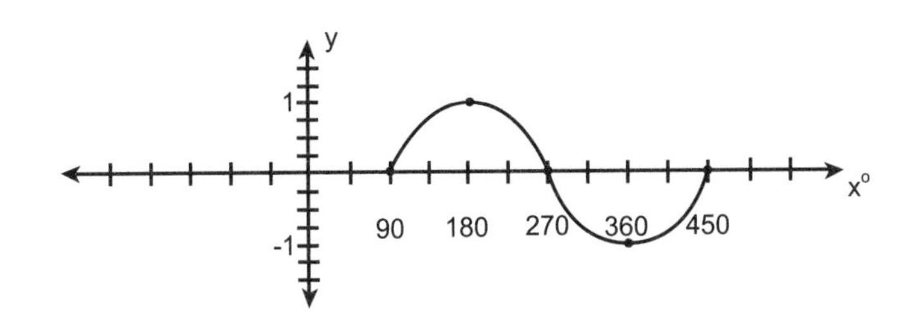

1. $y = \cos(x - 90°)$

x°		y
	0	
	90	
	180	
	270	
	360	

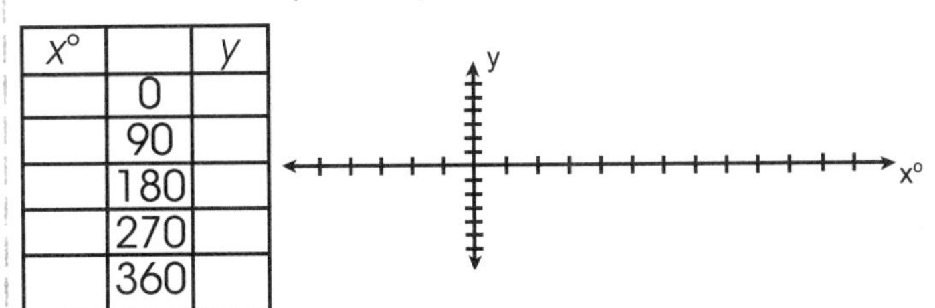

2. $y = \sin(x + 90°)$

x°		y
	0	
	90	
	180	
	270	
	360	

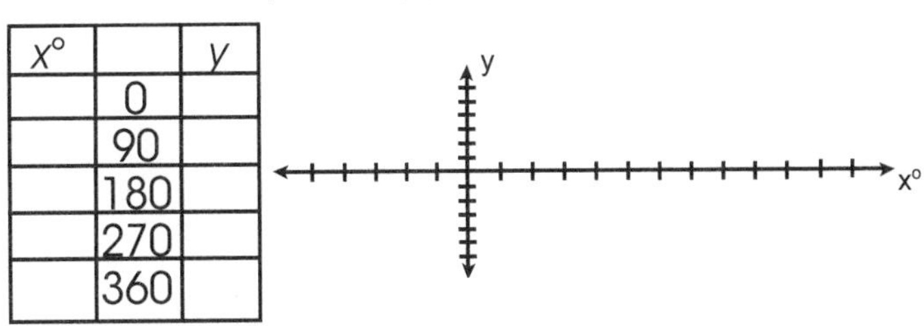

3. $y = \sin(x - 180°)$

x°		y
	0	
	90	
	180	
	270	
	360	

4. $y = \cos(x + 45°)$

x°		y
	0	
	90	
	180	
	270	
	360	

5. $y = \sin(x - 270°)$

x°		y
	0	
	90	
	180	
	270	
	360	

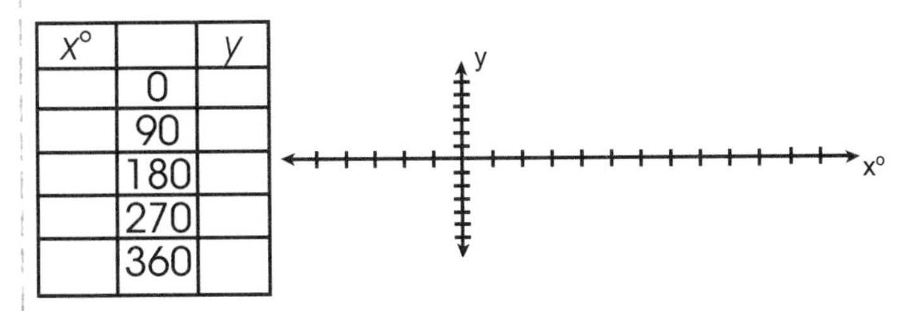

6. $y = \cos(x + 30°)$

x°		y
	0	
	90	
	180	
	270	
	360	

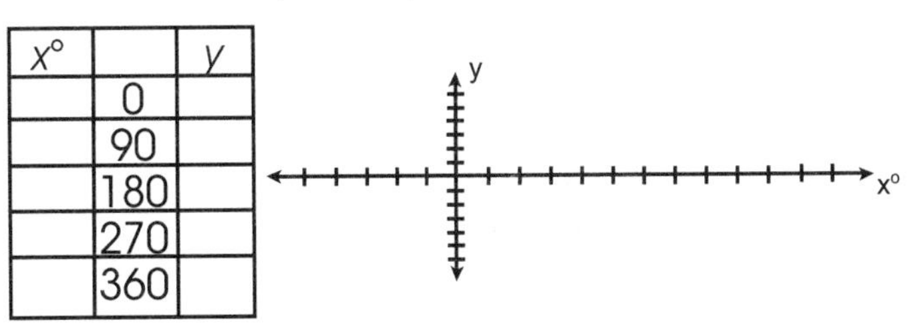

Name________________________________

HSF-TF.B.6, HSF-TF.B.7

Graphing $y = c + \sin(x - d)$ or $y = c + \cos(x - d)$

In the general equation $y = c + \sin(x - d)$ or $y = c + \cos(x - d)$, the variable c is known as the vertical translation and the variable d is known as the horizontal translation. Now that we are familiar with the general shape of the sine and cosine functions, we will not use a table of values, but simply translate the origin graph from (0,0) to (d,c). Note that when both c and d are accounted for by translating the origin, the function simplifies to $y = \sin x$ or $y = \cos x$.

Graph each function.

Example: $y = 2 + \sin(x + 90)$

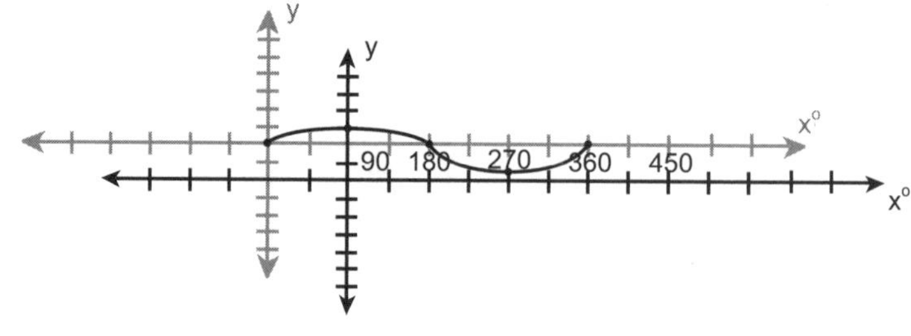

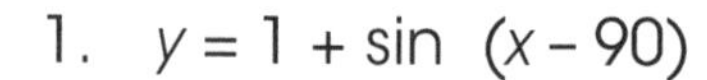

1. $y = 1 + \sin(x - 90)$

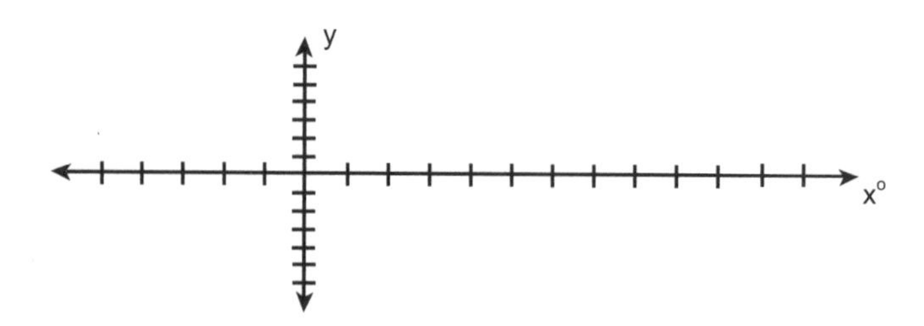

2. $y = 2 + \cos(x - 45)$

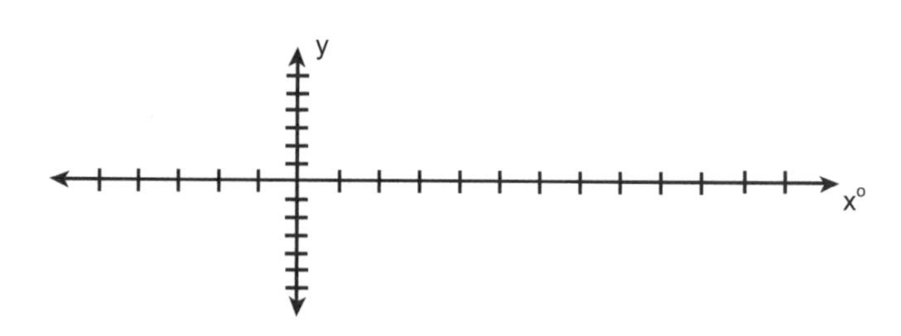

3. $y = {}^{-}1 + \cos(x + 90)$

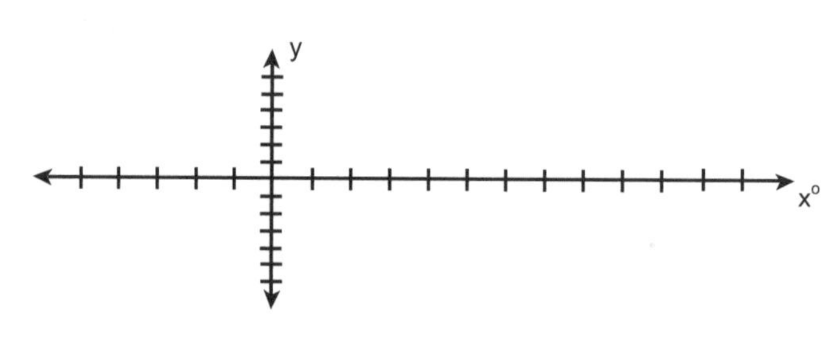

4. $y = {}^{-}2 + \sin(x + 180)$

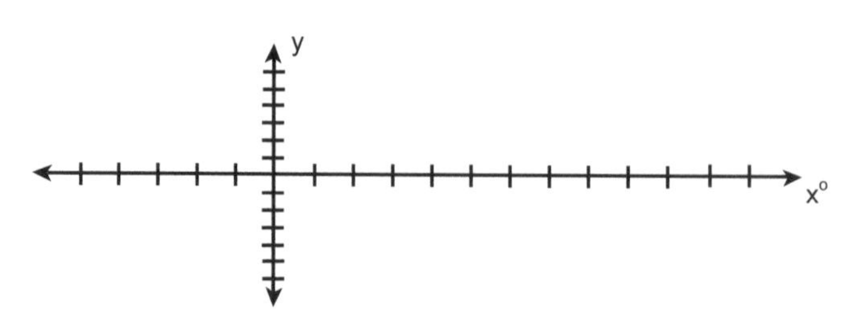

Name____________________________________ HSF-TF.B.6, HSF-TF.B.7

Graphing $y = c + a \sin b(x - d)$ or $y = c + a \cos b(x - d)$

In the general equation $y = c + a \sin b(x - d)$ or $y = c + a \cos b(x - d)$...

$|a|$ is the amplitude.

$\frac{360}{b}$ is the period.

c is the vertical translation.
d is the horizontal translation.

To graph a complex equation of this type, first translate the origin of the graph from (0, 0) to (d, c) which will reduce the equation to $y = a \sin bx$ or $y = a \cos bx$. Then, use a table of values with three columns (x, bx, and y) to graph the equation on the translated $x - y$ plane.

Example: $y = 1 + 3 \sin 2(x + 90)$

a. The first step is to translate the origin of the coordinate system from (0, 0) to ($^-90$, 1).

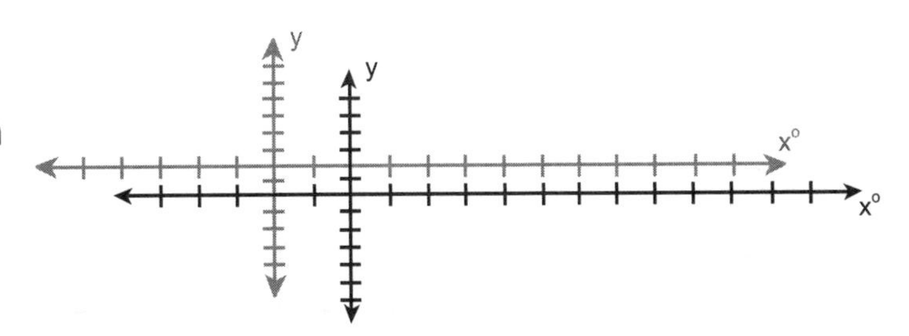

b. In translating the origin, the equation is reduced to $y = 3 \sin 2x$. The amplitude is 3, the period is 180 degrees.

c. After completing the table of values, graph the first and last columns only onto the translated coordinate system.

$x°$	$2x°$	y
0	0	0
45	90	3
90	180	0
135	270	$^-3$
180	360	0

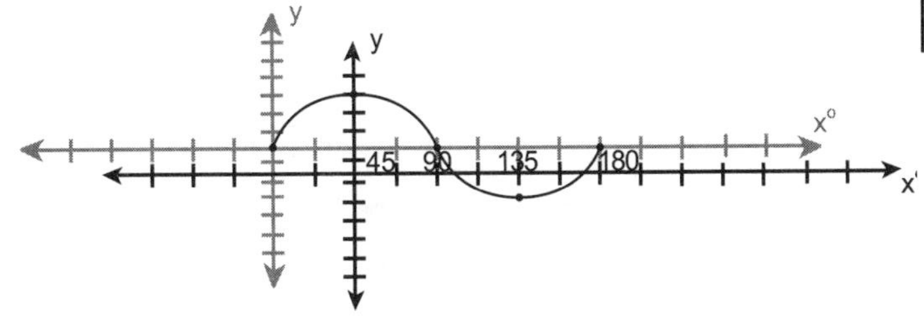

Name____

HSF-TF.B.6, HSF-TF.B.7

More Graphing $y = c + a \sin b\,(x - d)$ or $y = c + a \cos b\,(x - d)$

Graph the following equations. Write the units on both axes.

1. $y = {}^{-}2 - 1 \cos 3\,(x - 90)$

$x°$		y

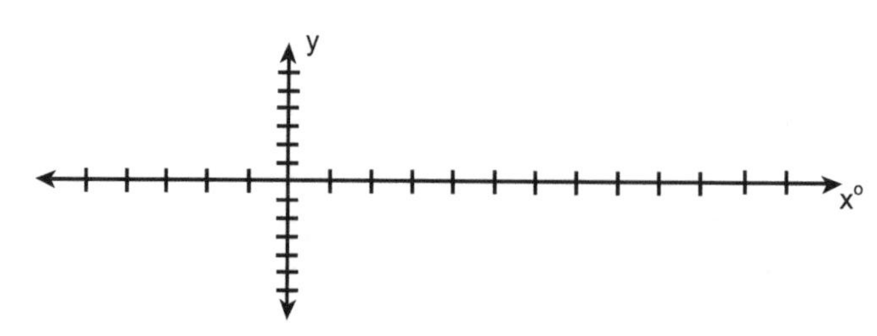

2. $y = {}^{-}1 + \frac{1}{2} \cos 2\,(x - 180)$

$x°$		y

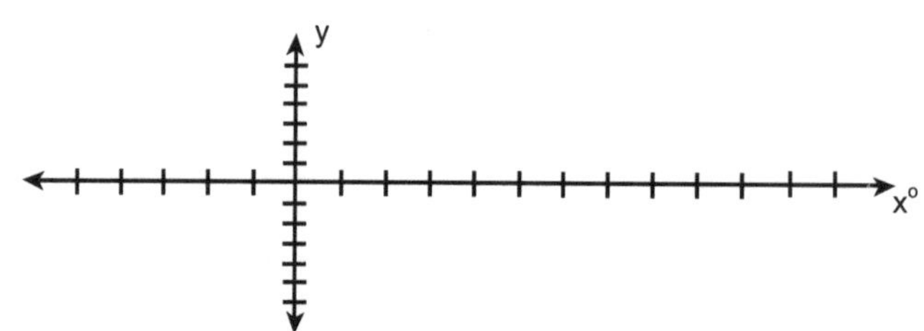

3. $y = 3 + 3 \sin (x + 90)$

$x°$		y

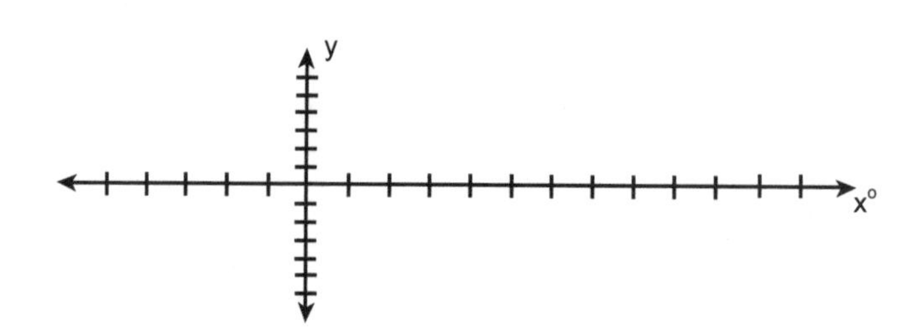

4. $y = {}^{-}6 - 2 \sin \frac{1}{2}\,(x + 45)$

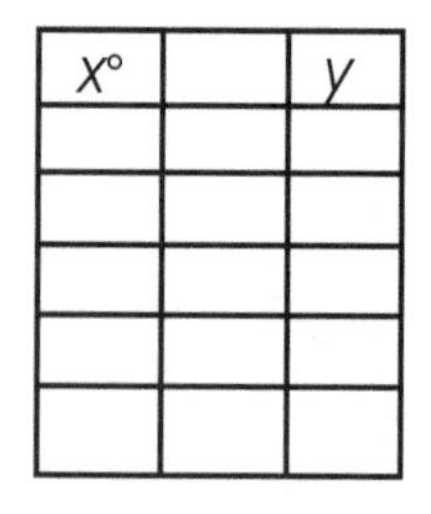

$x°$		y

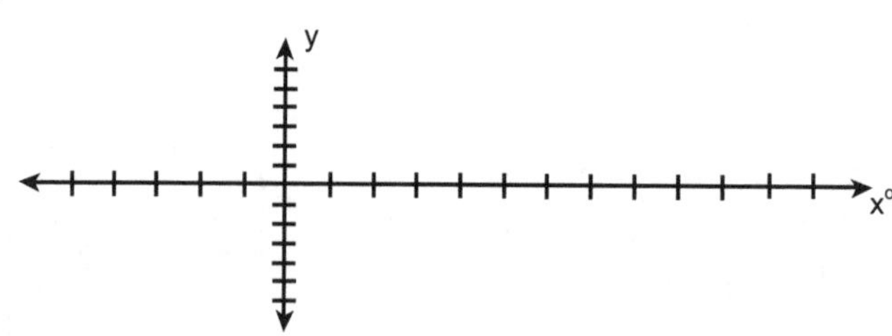

5. $y = 2 + 3 \cos 4\,(x - 45)$

$x°$		y

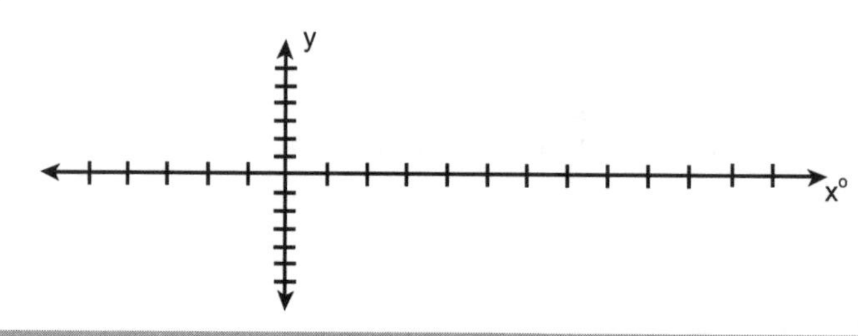

Name_______________________________________ HSG-SRT.D.10, HSG-SRT.D.11

Law of Sines

Given the triangle ABC with lengths *a*, *b*, *c* and angles A, B, and C then,

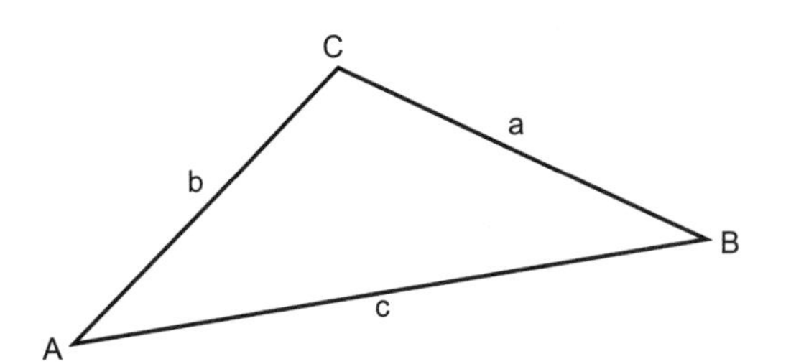

1. $\frac{\sin A}{a} = \frac{\sin B}{b}$
2. $\frac{\sin C}{c} = \frac{\sin B}{b}$
3. $\frac{\sin A}{a} = \frac{\sin C}{c}$

Note: *a* is the length of the side opposite angle *A*.

Solve for all missing sides and angles in each triangle. Round sides to nearest tenth and angles to nearest degree.

Example:

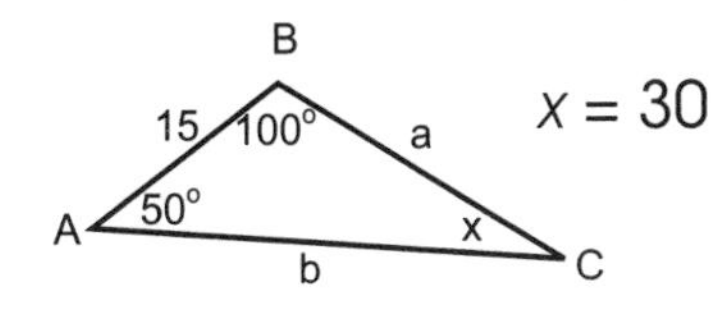

$100 + 50 + x = 180$

$x = 30$

$\frac{\sin 50}{a} = \frac{\sin 30}{15}$

$15 \sin 50 = a \sin 30$

$a = \frac{15 \sin 50}{\sin 30} = 22.981$

$a = 22.9$

$\frac{\sin 100}{b} = \frac{\sin 30}{15}$

$15 \sin 100 = b \sin 30$

$b = \frac{15 \sin 100}{\sin 30} = 29.50$

$b = 29.5$

Note: The law of sines works for all triangles.

1. 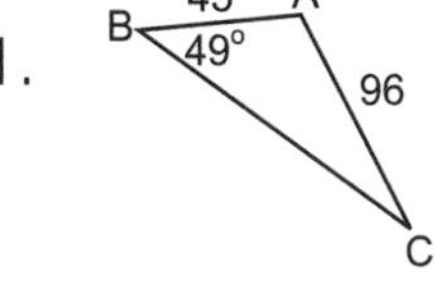

2. 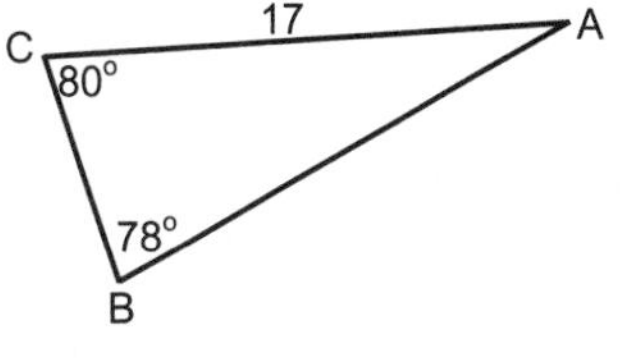

3. 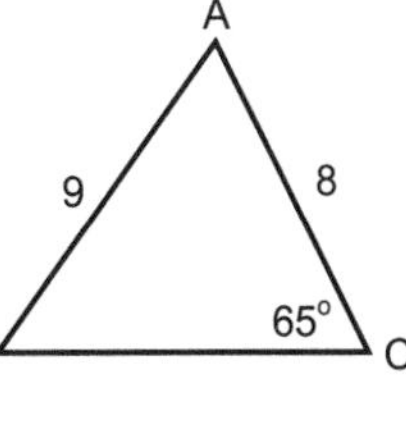

4. 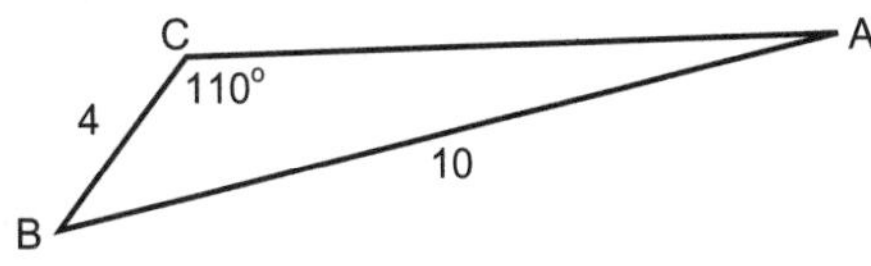

5. 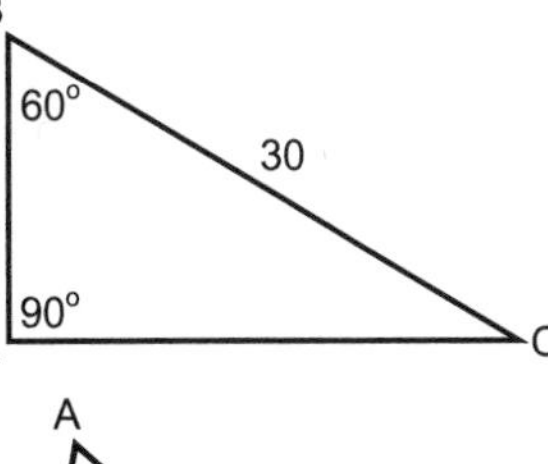

6. 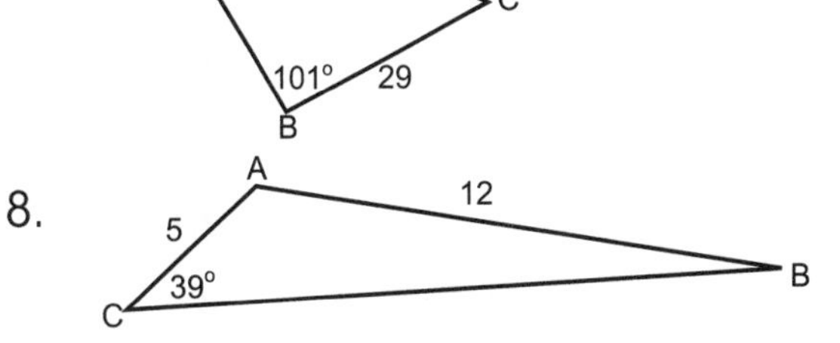

7. 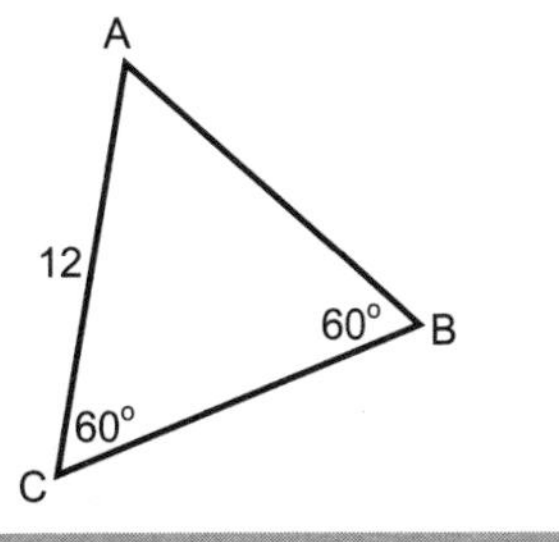

8.

Name________________________________

Law of Cosines

Given the triangle ABC with lengths *a*, *b*, *c* and angles A, B, and C then,

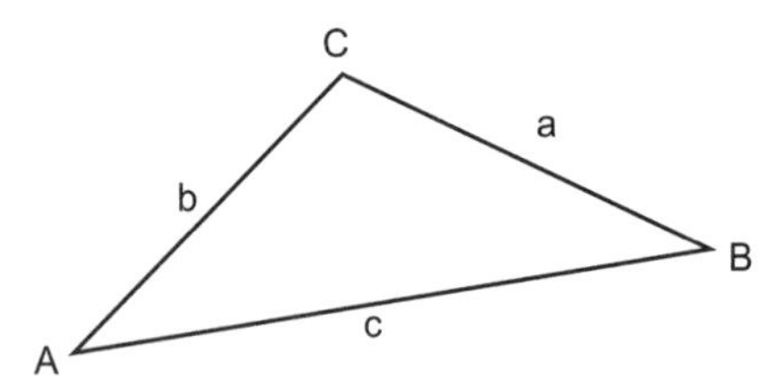

1. $a^2 = b^2 + c^2 - 2bc \cos A$
2. $b^2 = a^2 + c^2 - 2ac \cos B$
3. $c^2 = a^2 + b^2 - 2ab \cos C$

Note: *a* is the length of the side opposite angle *A*.

The law of cosines is significantly more difficult to use than that of the law of sines. Consider two cases: The first is solving for side *a* of a triangle, the second is solving for angle A.

1. $a^2 = \sqrt{b^2 + c^2 - 2bc \cos A}$
2. $A = \cos^{-1}\left(\frac{a^2 - b^2 - c^2}{^{-}2bc}\right)$

Solve for *x* on each triangle. Round length to nearest tenth and angle to the nearest degree.

Example:

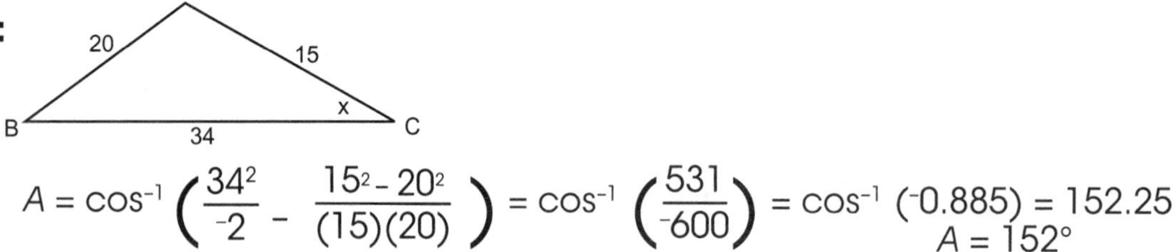

$$A = \cos^{-1}\left(\frac{34^2}{^{-}2} - \frac{15^2 - 20^2}{(15)(20)}\right) = \cos^{-1}\left(\frac{531}{^{-}600}\right) = \cos^{-1}(^{-}0.885) = 152.25$$
$$A = 152°$$

Note: The law of cosines works for all triangles.

1. B 45 A x° 60 96 C

2.

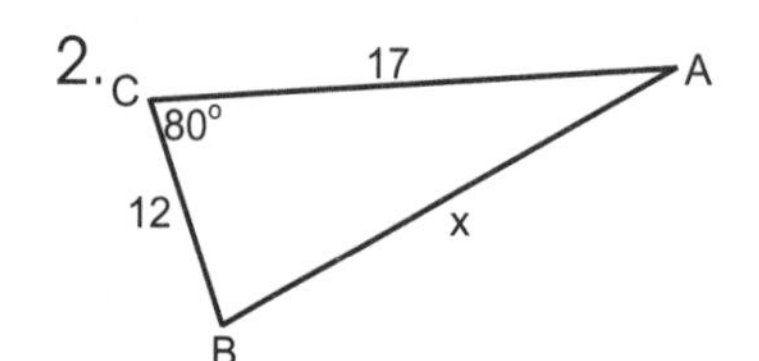

3.

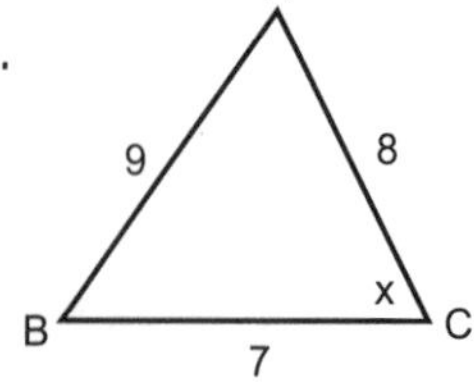

4.

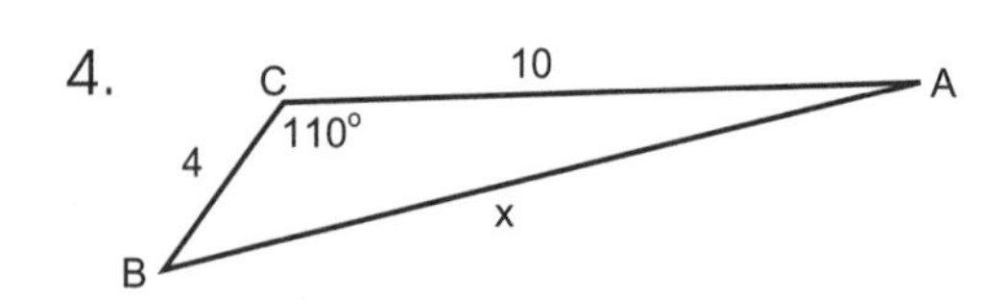

5.

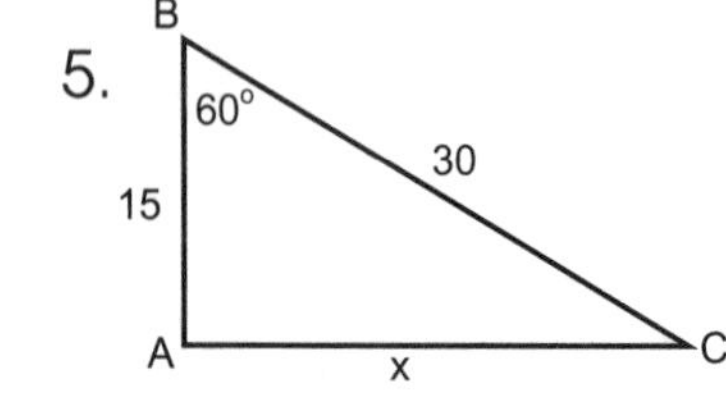

6.

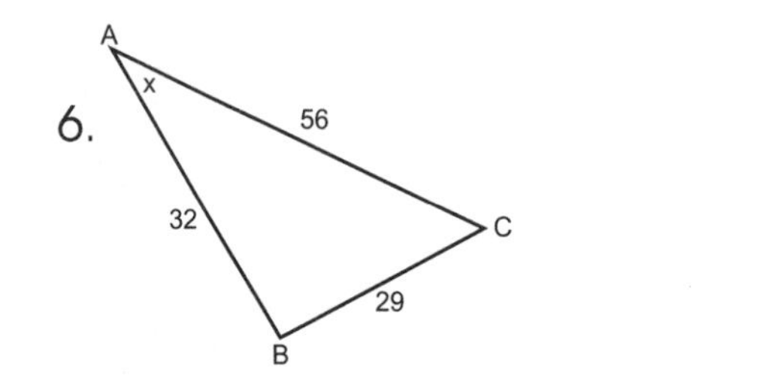

Name________________________

HSG-SRT.D.10, HSG-SRT.D.11

Problem Solving: The Law of Sines and the Law of Cosines

Draw a picture and solve.

1. Juan and Romelia are standing at the seashore 10 miles apart. The coastline is a straight line between them. Both can see the same ship in the water. The angle between the coastline and the line between the ship and Juan is 35 degrees. The angle between the coastline and the line between the ship and Romelia is 45 degrees. How far is the ship from Juan?

2. Jack is on one side of a 200-foot-wide canyon and Jill is on the other. Jack and Jill can both see the trail guide at an angle of depression of 60 degrees. How far are they from the trail guide?

3. Tom, Dick, and Harry are camping in their tents. If the distance between Tom and Dick is 153 feet, the distance between Tom and Harry is 201 feet, and the distance between Dick and Harry is 175 feet, what is the angle between Dick, Harry, and Tom?

4. Three boats are at sea: Jenny one (J1), Jenny two (J2), and Jenny three (J3). The crew of J1 can see both J2 and J3. The angle between the line of sight to J2 and the line of sight to J3 is 45 degrees. If the distance between J1 and J2 is 2 miles and the distance between J1 and J3 is 4 miles, what is the distance between J2 and J3?

5. Airplane A is flying directly toward the airport which is 20 miles away. The pilot notices airplane B 45 degrees to her right. Airplane B is also flying directly toward the airport. The pilot of airplane B calculates that airplane A is 50 degrees to his left. Based on that information, how far is airplane B from the airport?

Name________________________________ HSF-LE.A.2, HSF-IF.C.7a

Graphing Parabolas

General Equation: $y = a(x - h)^2 + k$ or $x = a(y - k)^2 + h$

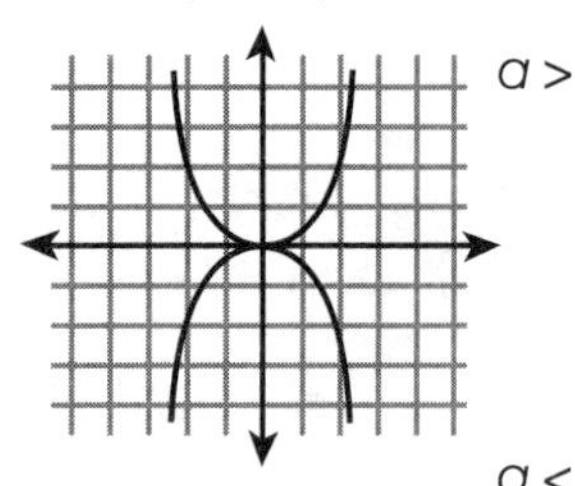
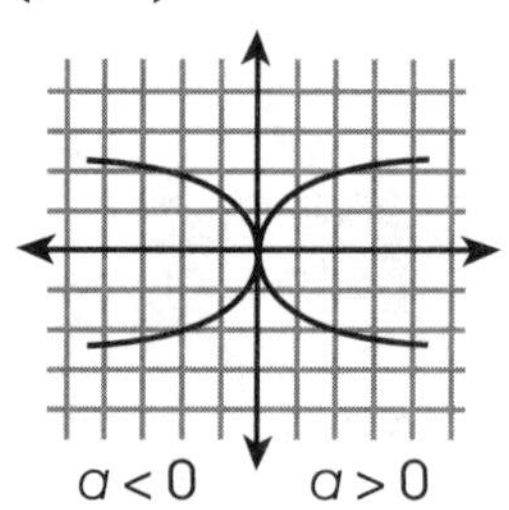

The vertex of the parabola is (h, k).

Graph each parabola and label its vertex.

Example: $y = 3(x - 3)^2 + 2$

x	y
2	5
4	5

Choose one x value on either side of vertex.
vertex (3, 2)

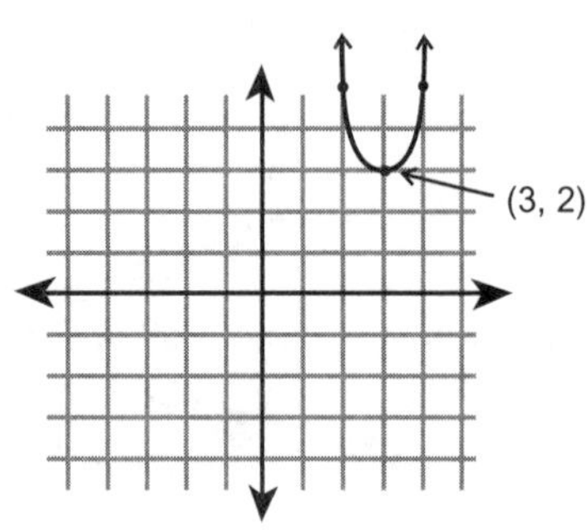

1. $x = (y + 2)^2 + 4$

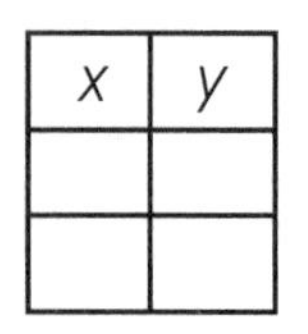

x	y

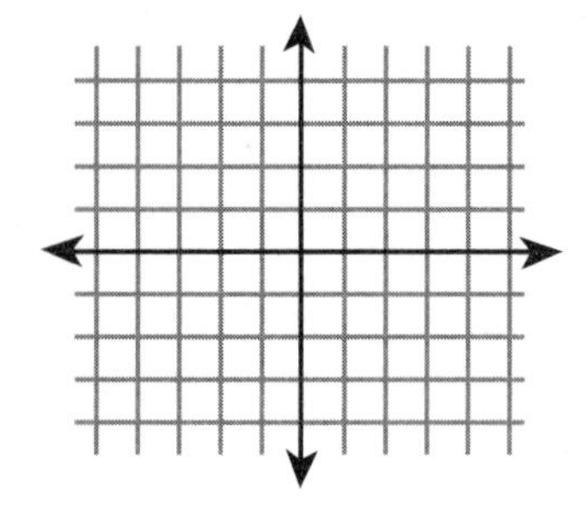

2. $x = {}^-2(y - 2)^2 + 3$

x	y

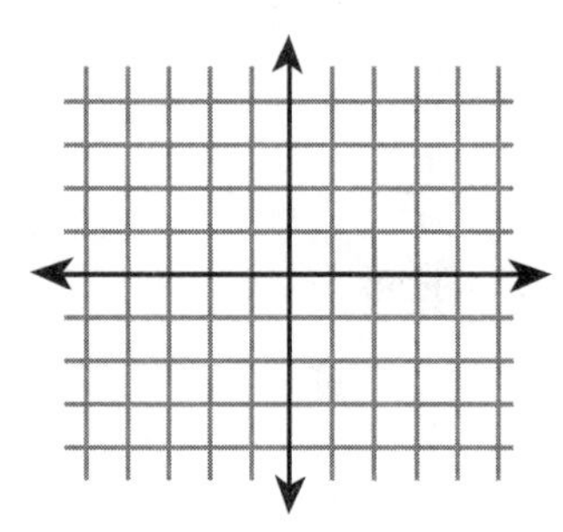

3. $y = -(x + 3)^2 - 1$

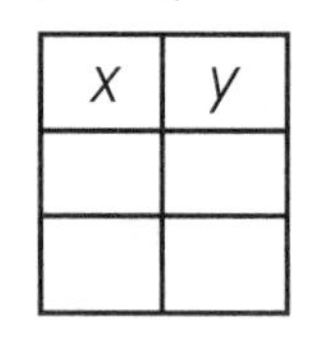

x	y

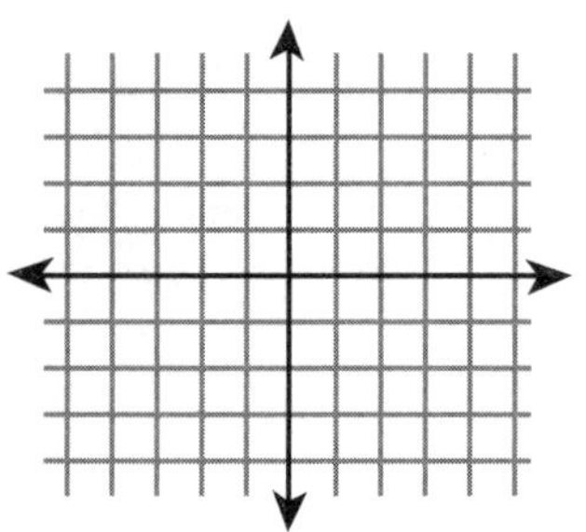

4. $y = \frac{1}{3}(x + 5)^2$

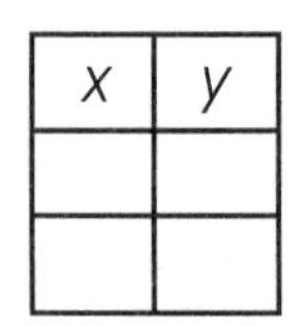

x	y

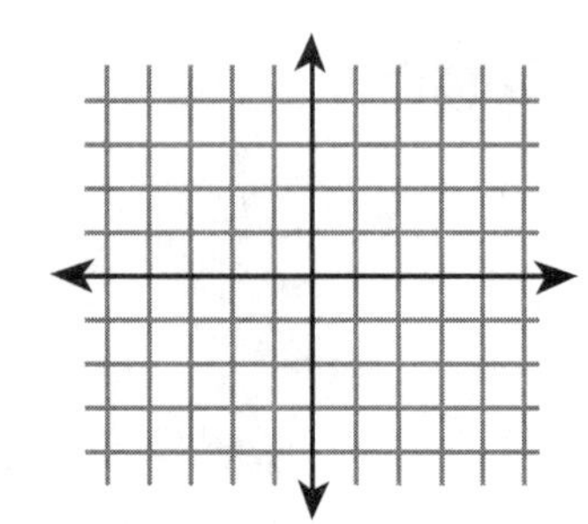

5. $x = \frac{1}{10}(y + 2)^2 + 2$

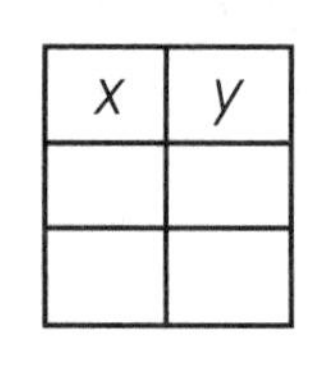

x	y

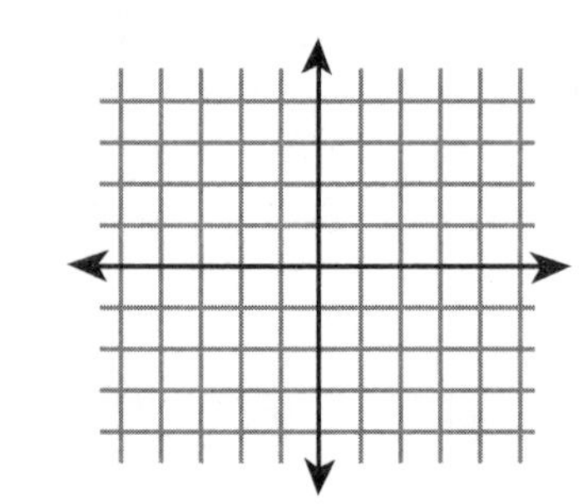

Name____________________________________ HSF-IF.C.8a, HSF-BF.A.2

Completing the Square

Parabolas as well as other conic sections are not always given in their general form. To put a conic section in its general form it is sometimes necessary to "complete the square."

Example: $y = 3x^2 - 18x - 10$

To complete the square. . .

1. Isolate the x terms. $y + 10 = 3x^2 - 18x$
2. Divide by the x^2 coefficient. $\frac{(y + 10)}{3} = 1x^2 - 6x$
3. Divide the coefficient of the x term by two, then square it and add the product to both sides of the equation. $\frac{(y + 10)}{3} + 9 = 1x^2 - 6x + 9$
4. Factor the right side of the equation. $\frac{(y + 10)}{3} + 9 = (x - 3)(x - 3)$

   $3\left[\frac{(y + 10)}{3}\right] = 3(x - 3)^2$
5. Solve for y. $y + 10 + 27 = 3(x - 3)^2$

   $y + 37 = 3(x - 3)^2$

   $y = 3(x - 3)^2 - 37$

vertex of parabola $(3, {}^-37)$

Complete the square and name the vertex for each parabola.

1. $y = 2x^2 - 4x + 8$
2. $y = {}^-3x^2 - 12x - 13$
3. $y = \frac{1}{3}x^2 - 2x + 3$
4. $y = \frac{1}{5}x^2 - \frac{2}{5}x + \frac{11}{5}$
5. $x = y^2 + 10y - 6$
6. $x = y^2 - 10y + 35$
7. $x = 5y^2 + 40y + 77$
8. $x = \frac{1}{2}y^2 - \frac{3}{2}y - \frac{1}{4}$

Name____________________

Equations for Circles

General equation: $(x - h)^2 + (y - k)^2 = r^2$

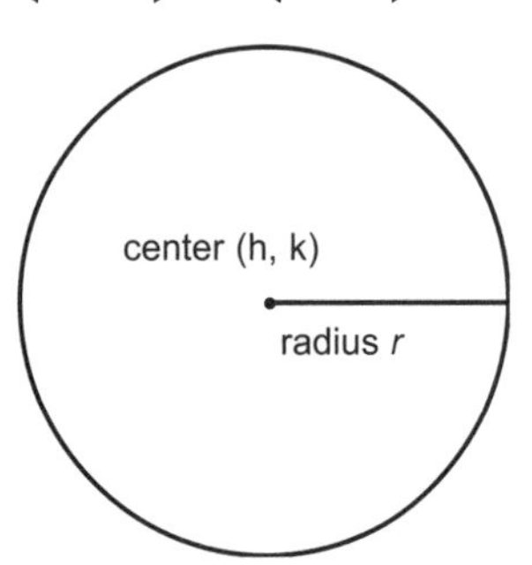

Given the equation for a circle, identify its center and its radius.

Example: $(x - 2)^2 + (y - 3)^2 = 25$
center (2, 3)
radius = 5

1. $(x - 4)^2 + (y + 10)^2 = 144$

2. $x^2 + (y - 7)^2 = 49$

3. $x^2 + y^2 = 1$

4. $(x + 3)^2 + (y + 11)^2 = 15$

5. $(x - 15)^2 + y^2 = 10$

Given the center and the radius of a circle, write the equation describing the circle.

Example: (0, 4), $r = 9$
$(x - 0)^2 + (y - 4)^2 = 81$
$x^2 + (y - 4)^2 = 81$

1. (0, 0), $r = 8$

2. ($^-2$, 3), $r = 2$

3. ($^-7$, $^-18$), $r = 14$

4. (12, 9), $r = 1$

5. (10, 0), $r = 22$

Name____________________________________

HSF-IF.C.7a

Graphing Circles

Graph each circle and label its center and radius.

Example: $(x-2)^2+(y+5)^2=4$
center (2, ⁻5)
radius = 2

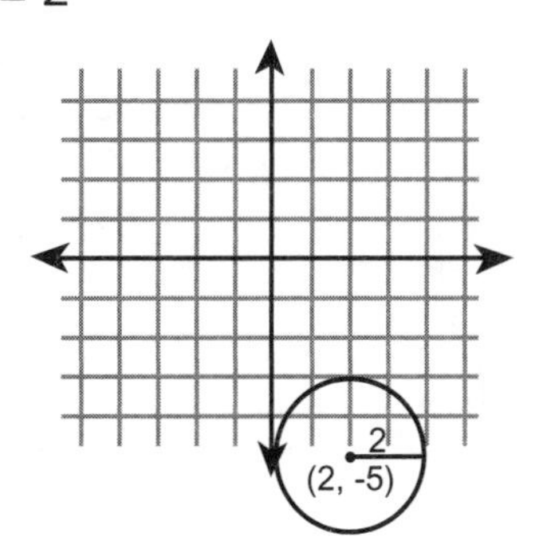

1. $x^2+(y-3)^2=16$

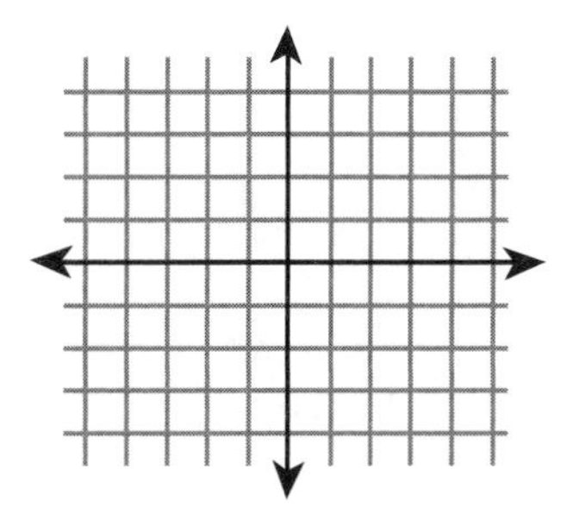

2. $x^2+y^2=64$

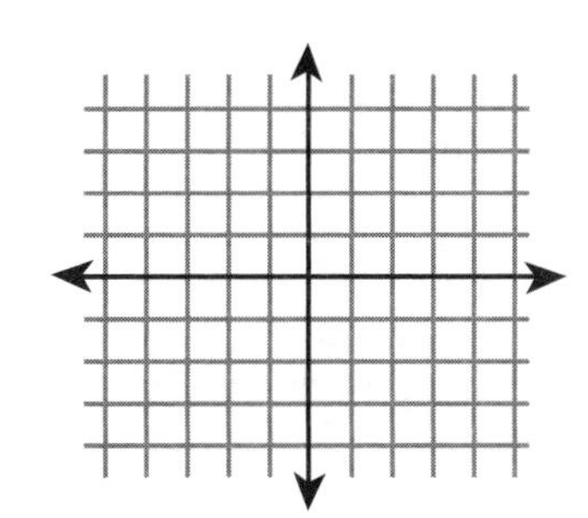

3. $(x-1)^2+(y+1)^2=1$

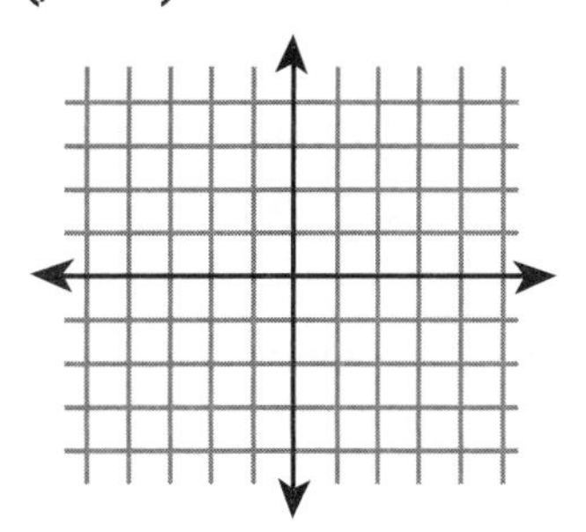

4. $(x-7)^2+(y-2)^2=25$

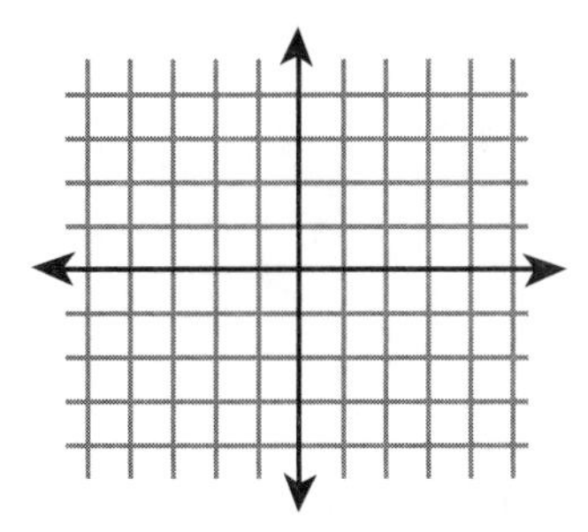

5. $(x+4)^2+y^2=9$

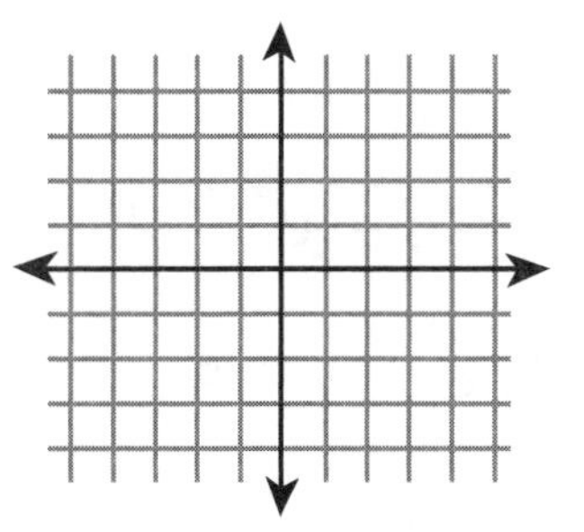

6. $x^2+(y-12)^2=20$

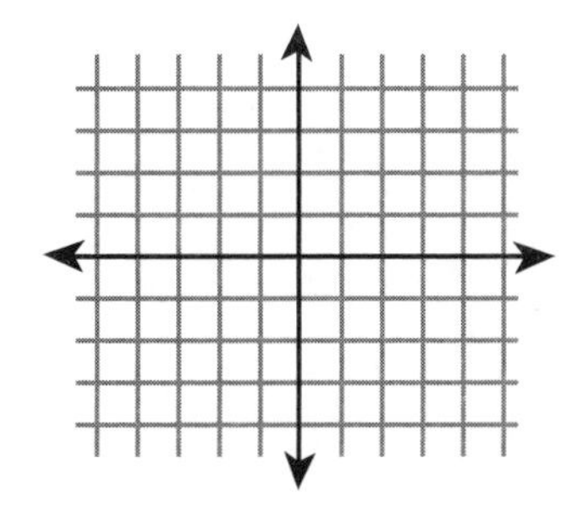

7. $(x+6)^2+(y+9)^2=15$

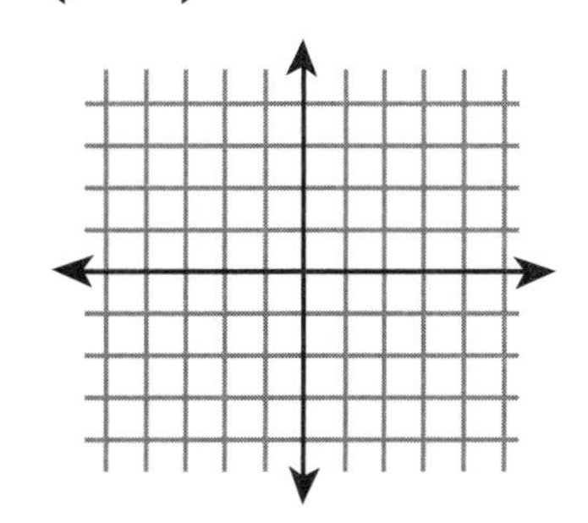

Name________________________________

HSF-LE.A.2, HSF-BF.A.2

Graphing an Ellipse with Center (0, 0)

Two different types of ellipses may have the center (0, 0). The first type has its two foci on the *x*-axis, and the second type has its two foci on the *y*-axis. The axis which contains the foci is called the major axis. The other axis is called the minor axis.

Type one: *x*-axis as the major axis

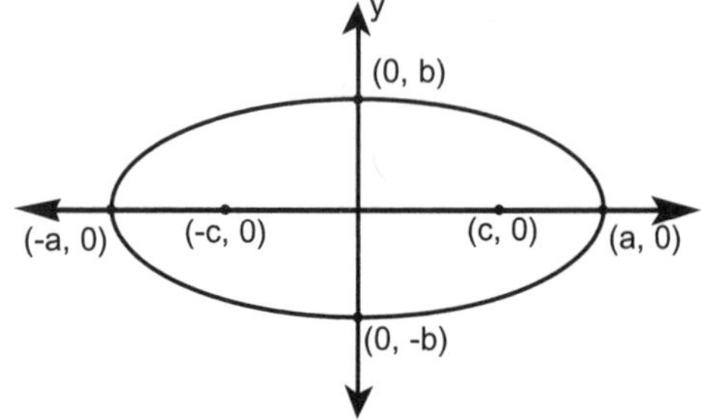

General equation: $\frac{x^2}{a^2} + \frac{y^2}{b^2} = 1$

The vertices of the ellipse are $(-a, 0)$, $(a, 0)$, $(0, b)$, $(0, -b)$.

The foci of the ellipse are $(c, 0)$, $(-c, 0)$.

The equation relating a, b, and c is $c^2 = a^2 - b^2$.

Note: $c^2 = a^2 - b^2$ looks like the Pythagorean theorem and it is derived from it, but a is the hypotenuse rather than c.

Type two: *y*-axis as the major axis

General equation: $\frac{x^2}{b^2} + \frac{y^2}{a^2} = 1$

The vertices of the ellipse are $(0, -a)$, $(0, a)$, $(b, 0)$, $(-b, 0)$.

The foci of the ellipse are $(0, -c)$, $(0, c)$.

The equation relating a, b, and c is $c^2 = a^2 - b^2$.

Note: In the general equation, the variable a^2 is always the larger of the two denominators and is under the variable of the major axis.

Name________________________________ HSF-IF.C.7a

More Graphing an Ellipse with Center (0, 0)

Graph each ellipse and label the four vertices and the foci.

Example: $\frac{x^2}{9} + \frac{y^2}{25} = 1$

$b^2 = 9$ so $b = 3$ and $a^2 = 25$ so $a = 5$
and $c^2 = a^2 - b^2$ therefore $c = \sqrt{a^2 - b^2}$
finally, $c = \sqrt{25 - 9}$ so $c = 4$

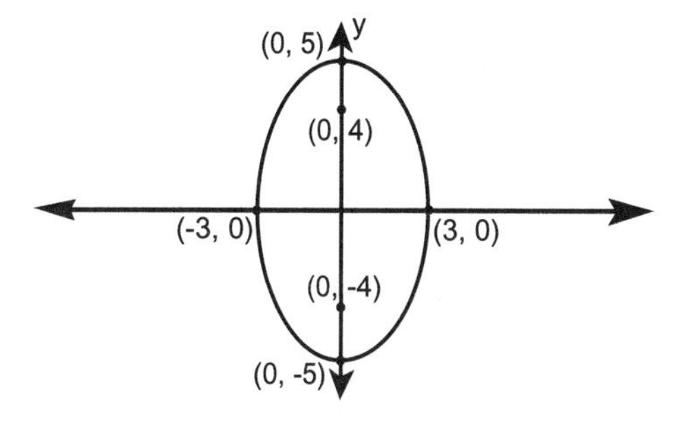

1. $\frac{x^2}{4} + \frac{y^2}{25} = 1$

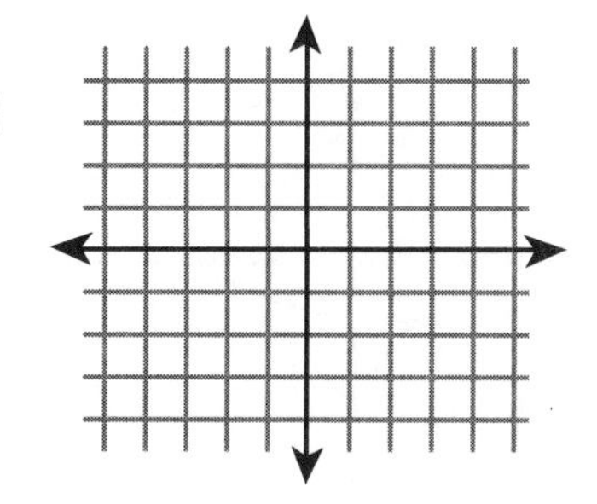

2. $\frac{x^2}{64} + y^2 = 1$

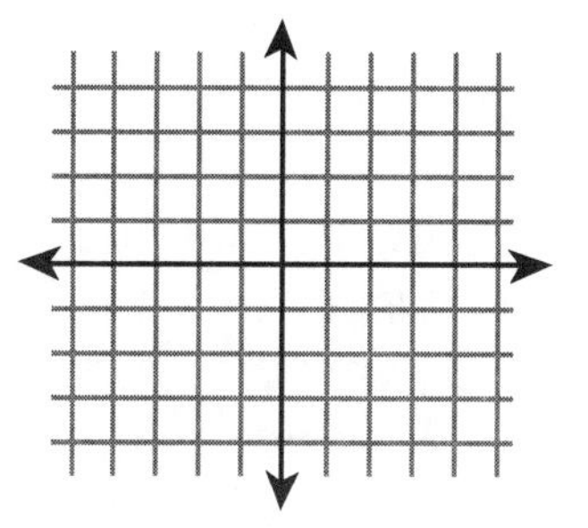

3. $\frac{x^2}{100} + \frac{y^2}{36} = 1$

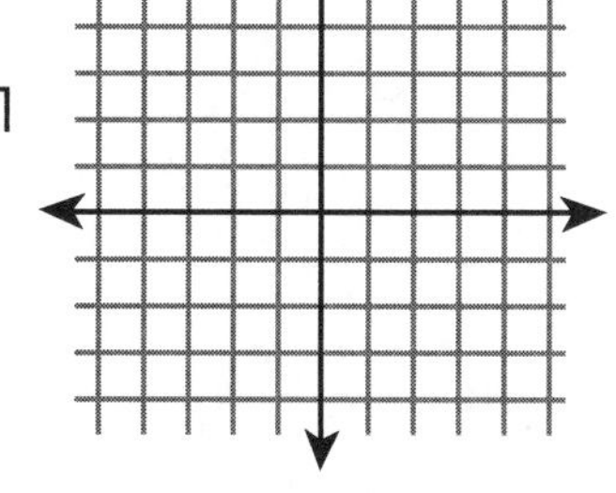

4. $\frac{x^2}{49} + \frac{y^2}{144} = 1$

5. $x^2 + \frac{y^2}{25} = 1$

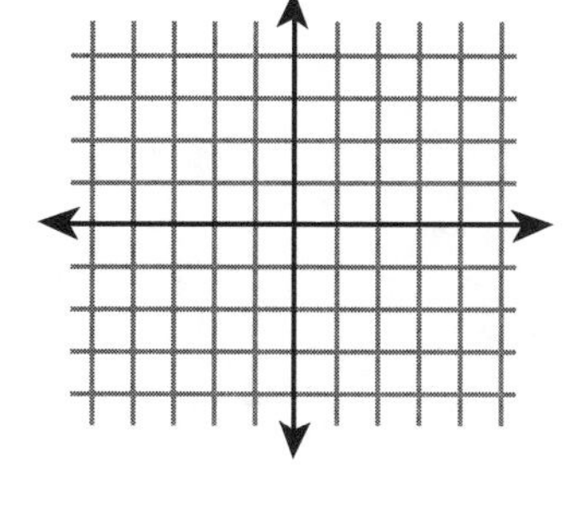

6. $\frac{x^2}{16} + \frac{y^2}{121} = 1$

Name________________________________ HSF-IF.C.7a, HSF-BF.A.2

Graphing Ellipses with Center (h, k)

There are two types of ellipses with center (h, k). The first type has its two foci on a horizontal major axis, and the second type has its two foci on a vertical major axis.

Type one: horizontal major axis

General equation:

$$\frac{(x-h)^2}{a^2} + \frac{(y-k)^2}{b^2} = 1$$

Type two: vertical major axis

General equation:

$$\frac{(x-h)^2}{b^2} + \frac{(y-k)^2}{a^2} = 1$$

Graph each ellipse, labeling the four vertices and the center. Mark units on both axes.

Example: $\frac{(x-1)^2}{36} + \frac{(y-2)^2}{9} = 1$

center = ($^-1$, 2), $a = 6$, $b = 3$
Note: a^2 is under the variable x. The x-axis is a horizontal line, therefore, this equation has a horizontal major axis.

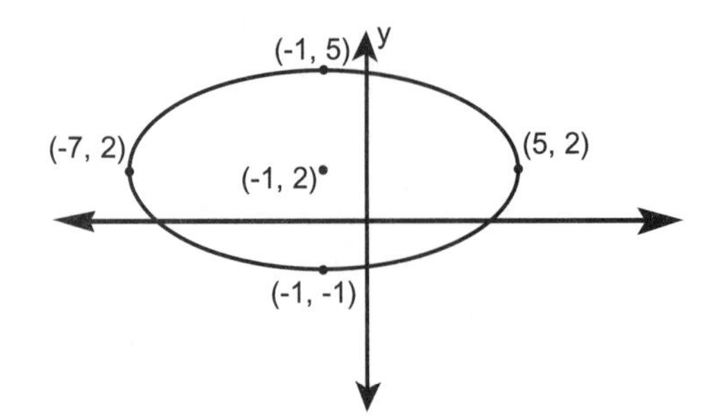

1. $\frac{(x-3)^2}{81} + \frac{(y-5)^2}{25} = 1$

2. $\frac{(x+8)^2}{64} + (y-6)^2 = 1$

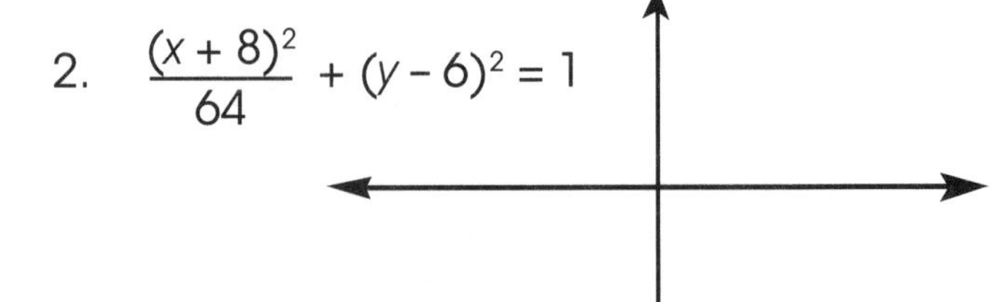

3. $\frac{x^2}{9} + \frac{y^2}{64} = 1$

4. $\frac{(x+2)^2}{49} + \frac{(y+3)^2}{255} = 1$

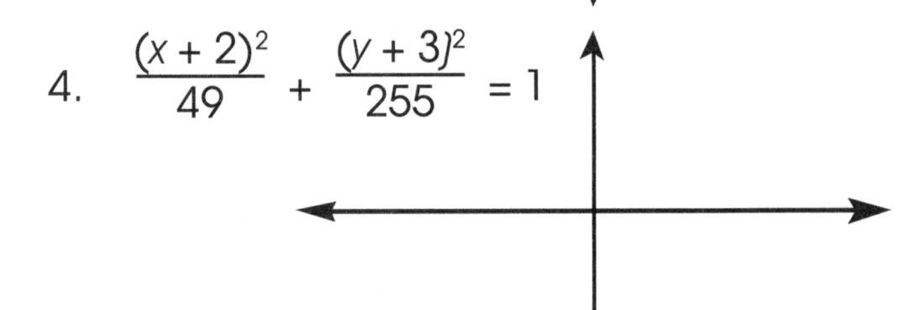

5. $\frac{(x+4)^2}{25} + \frac{(y-16)^2}{30} = 1$

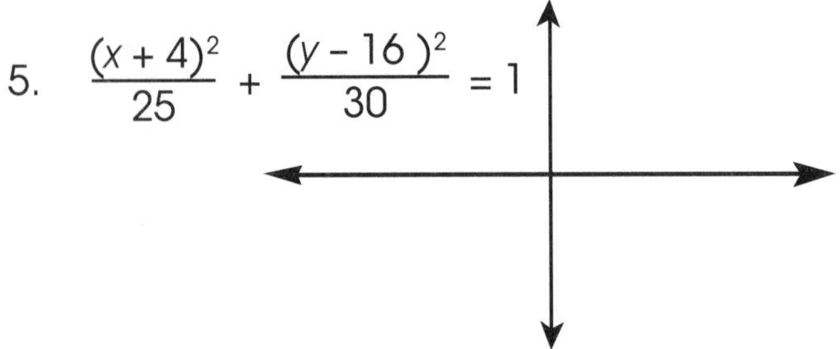

6. $\frac{(x-5)^2}{20} + \frac{(y-11)^2}{16} = 1$

Name________________________________ HSF-LE.A.2, HSF-BF.A.2

Graphing Hyperbolas with Center (0, 0)

There are two types of hyperbola with center (0, 0). The first type has its vertex and foci on the x-axis, and the second type has its vertex and foci on the y-axis.

Type one: vertex and foci on the x-axis

General equation: $\frac{x^2}{a^2} - \frac{y^2}{b^2} = 1$

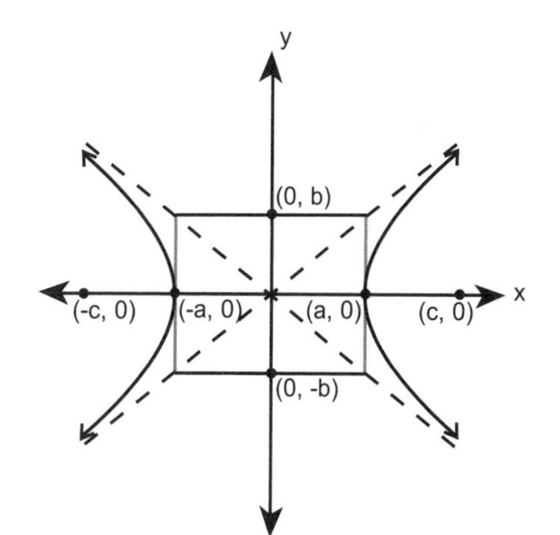

The vertices of the hyperbola are $(-a, 0)$, $(a, 0)$.

The foci of the hyperbola are $(c, 0)$, $(-c, 0)$.

The equation relating a, b, and c is $c^2 = a^2 + b^2$.

The two dashed lines are the asymptotes. The hyperbola gets very close to but does not touch either of these two dashed lines.

Note: The asymptotes are the diagonals of the quadrilateral with dimensions 2a by 2b.

Type two: vertex and foci on y-axis

General equation: $\frac{y^2}{a^2} - \frac{x^2}{b^2} = 1$

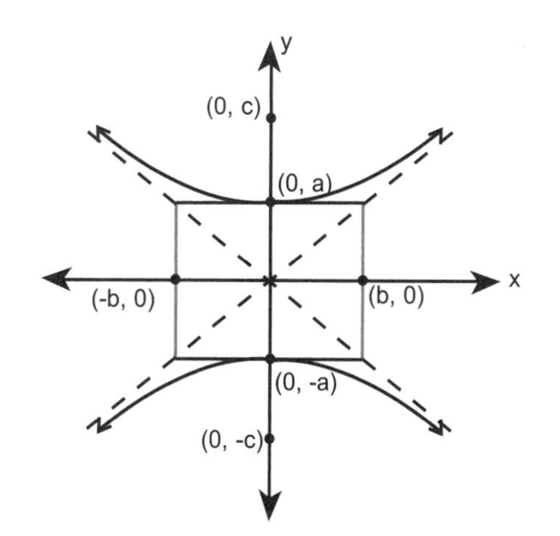

The vertices of the hyperbola are $(0, -a)$, $(0, a)$.

The foci of the hyperbola are $(0, -c)$, $(0, c)$.

The equation relating a, b, and c is $c^2 = a^2 + b^2$.

Note: The equations for hyperbolas and ellipses are identical except for a minus sign. In graphing either, the positive axis variable, x or y, is the axis which contains the vertex and the foci.

Name________________________________

HSF-IF.C.7a

More Graphing Hyperbolas with Center (0, 0)

Graph each hyperbola, labeling the vertices and foci.

Example: $\frac{x^2}{16} - \frac{y^2}{9} = 1$

$a^2 = 16$ so $a = 4$ $b^2 = 9$ so $b = 3$

$c^2 = a^2 + b^2$ therefore $c = \sqrt{a^2 + b^2}$

finally, $c = \sqrt{16 + 9}$ so $c = 5$

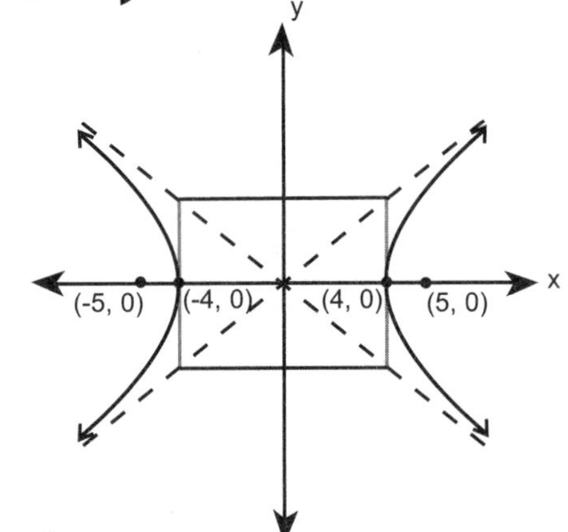

1. $\frac{x^2}{25} - \frac{y^2}{49} = 1$

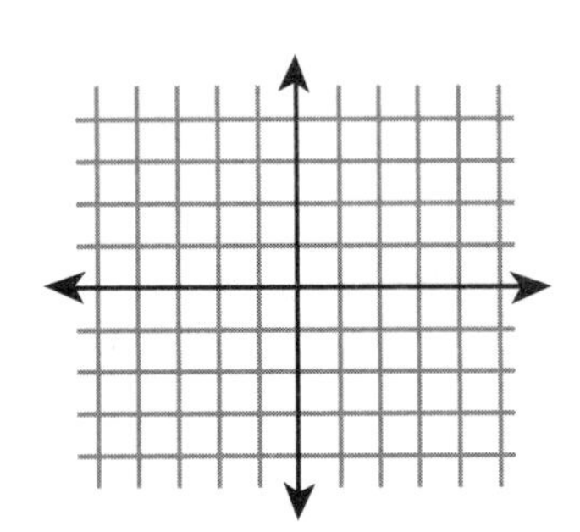

2. $\frac{y^2}{16} - \frac{x^2}{81} = 1$

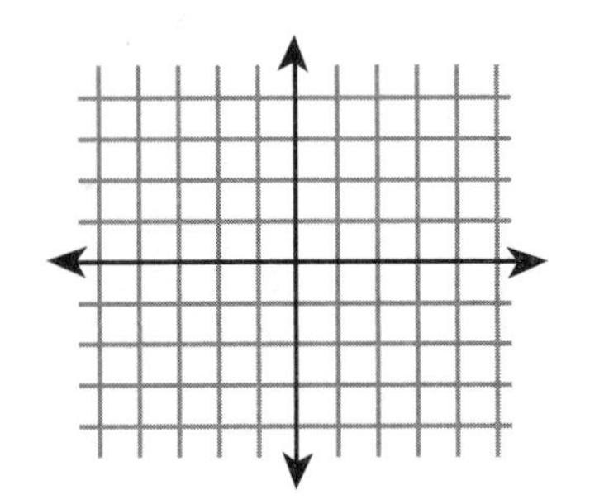

3. $\frac{y^2}{36} - \frac{x^2}{100} = 1$

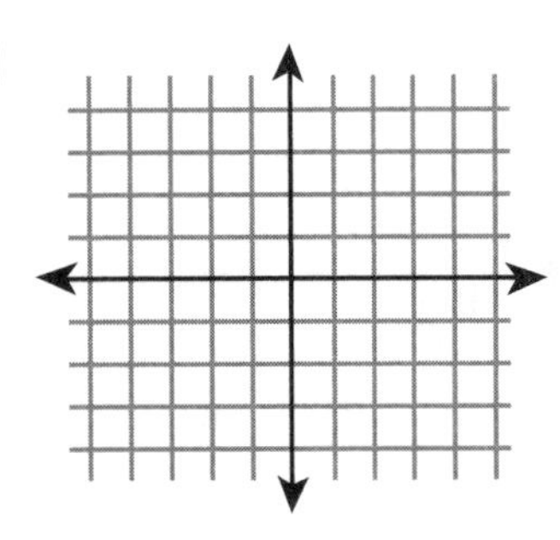

4. $\frac{x^2}{49} - \frac{y^2}{144} = 1$

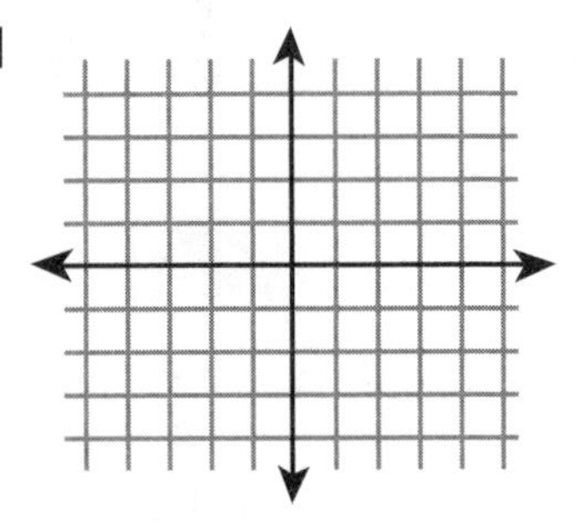

5. $x^2 - \frac{y^2}{25} = 1$

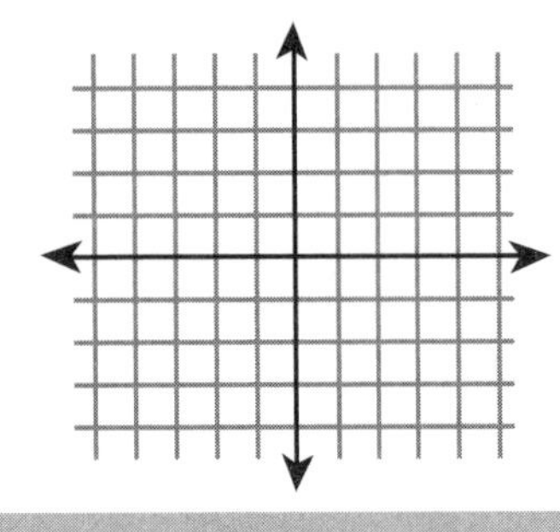

6. $\frac{y^2}{121} - \frac{x^2}{16} = 1$

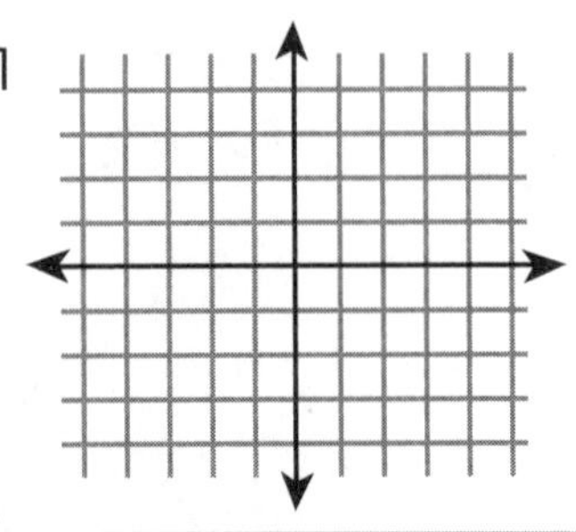

Name________________________________ HSF-IF.C.7a, HSF-BF.A.2

Graphing Hyperbolas with Center (*h*, *k*)

There are two types of hyperbola with center (h, k). In the first type, the vertex and foci are on the horizontal line $y = k$. In the second type, the vertex and foci are on the vertical line $x = h$.

Type one: horizontal line $y = k$.

General equation: $\frac{(x-h)^2}{a^2} - \frac{(y-k)^2}{b^2} = 1$

Type two: vertical line $x = h$.

General equation: $\frac{(y-k)^2}{a^2} - \frac{(x-h)^2}{b^2} = 1$

Graph each hyperbola, labeling the vertices and the center.

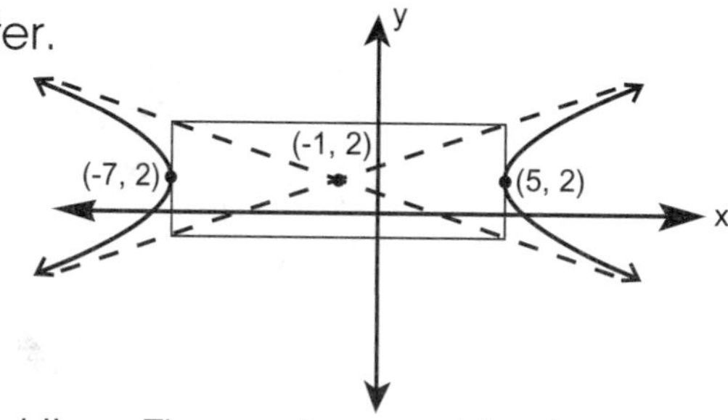

Example: $\frac{(x+1)}{36} - \frac{(y-2)}{9} = 1$

center = $(^-1, 2)$, $a = 6$, $b = 3$

Note: The variable x is positive since the x-axis is a horizontal line. The vertex and foci are on the horizontal line $y = k$.

1. $\frac{(x+3)^2}{4} - \frac{(y+5)^2}{121} = 1$

2. $(y+6)^2 - \frac{(x+8)^2}{16} = 1$

3. $\frac{x^2}{100} - \frac{y^2}{36} = 1$

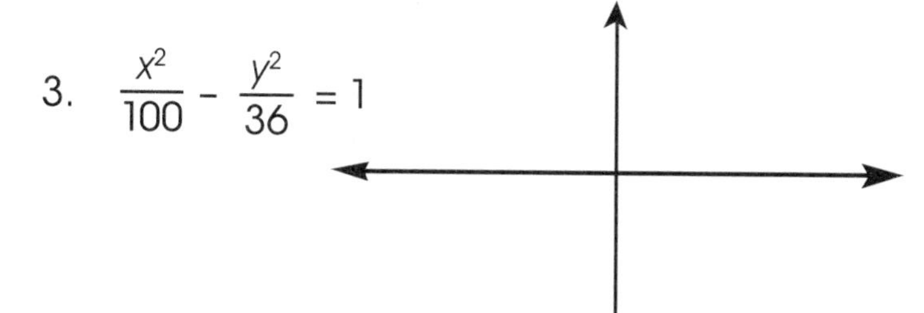

4. $\frac{(x+3)^2}{225} - \frac{(x+2)^2}{1} = 1$

5. $\frac{(y-16)^2}{36} - \frac{(x+4)^2}{25} = 1$

6. $\frac{(x-5)^2}{16} - \frac{(y-11)^2}{121} = 1$

Name____ HSF-IF.C.8a, HSF-BF.A.2

Identifying Different Types of Conic Sections

Complete the square, then identify the type of conic section and its center.

Example: $y = 16x^2 - 9y^2 - 32x + 72y - 272 = 0$

1. Isolate the x and y terms.
   $16x^2 - 32x - 9y^2 - 72y = 272$

2. Factor by the coefficients of the squared terms.
   $16(x^2 - 2x) - 9(y^2 - 8y) = 272 + 16(\quad) + {}^-9(\quad)$

3. Divide the coefficient of the x and y term by two, then square them, and add them to both sides of the equation.
   $16(x^2 - 2x + 1) - 9(y^2 - 8y + 16) = 272 + 16(1) + {}^-9(16)$

4. Factor the left side of the equation.
   $16(x - 1)^2 - 9(y - 4)^2 = 272 + 16(1) + {}^-9(16)$

5. Simplify.
   $16(x - 1)^2 - 9(y - 4)^2 = 272 + 16 + {}^-144$
   $16(x - 1)^2 - 9(y - 4)^2 = 144$

6. Divide by the product of the leading coefficients.
   $$\frac{16(x-1)^2}{144} - \frac{9(y-4)^2}{144} = \frac{144}{144}$$

   This is the equation of a hyperbola with center at (1, 4).
   $$\frac{(x-1)^2}{9} - \frac{(x-4)^2}{16} = 1$$

Write your answers in complete sentences.

1. $x^2 + y^2 - 4x + 6y + 4 = 0$

2. $9x^2 + 4y^2 + 54x + 8y + 49 = 0$

3. $25x^2 + y^2 - 300x + 8y + 891 = 0$

4. $16x^2 - y^2 + 96x + 8y + 112 = 0$

5. $x^2 + y^2 + 8x + 20y + 112 = 0$

Name________________________________

HSF-IF.C.7a

Graphing Conic Sections

Graph each conic section, name it, then label its vertices and center.

1. $x^2 + (y - 2)^2 = 25$

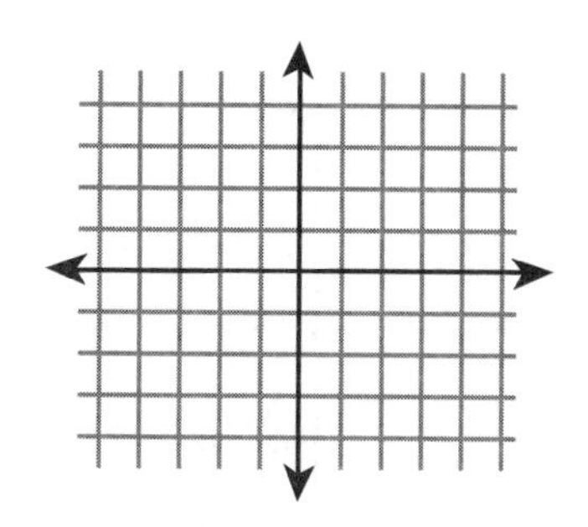

2. $\frac{x^2}{4} + \frac{(y - 3)^2}{16} = 1$

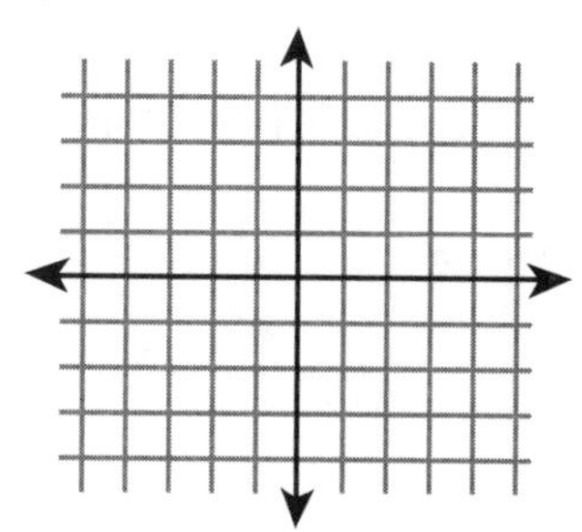

3. $\frac{(x - 1)^2}{16} - \frac{(y + 5)^2}{25} = 1$

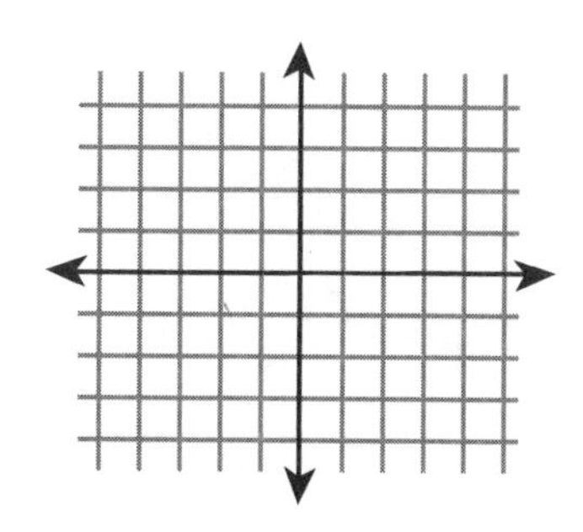

4. $y = (x + 2)^2 - 3$

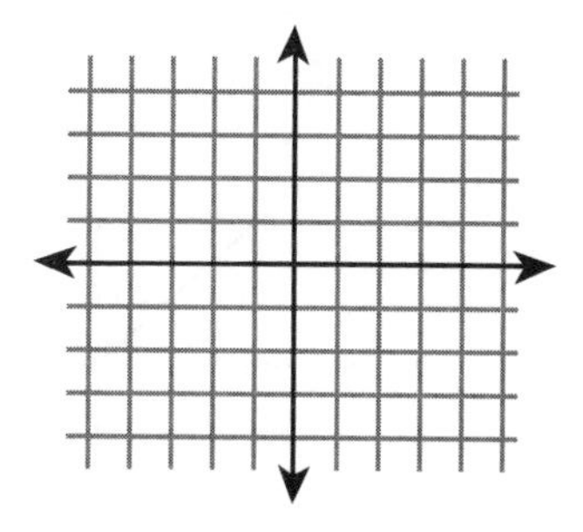

5. $x^2 + y^2 + 6x - 2y - 15 = 0$

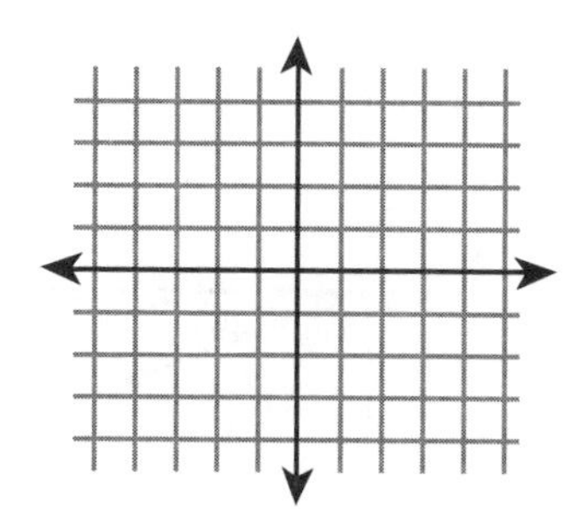

6. $x^2 + 4y^2 + 2x - 24y + 33 = 0$

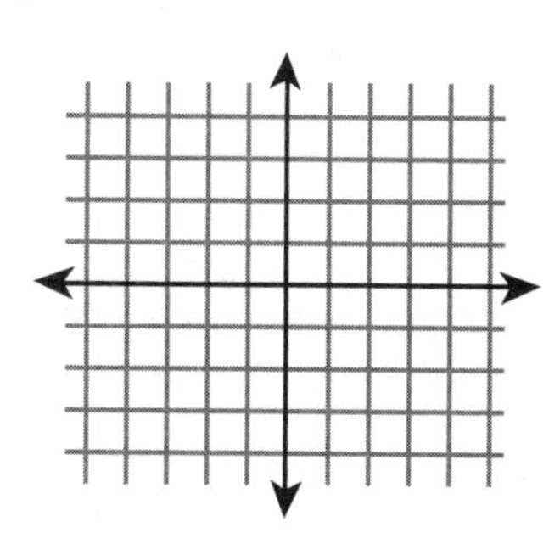

7. $y^2 - 4x^2 - 2y - 16x - 19 = 0$

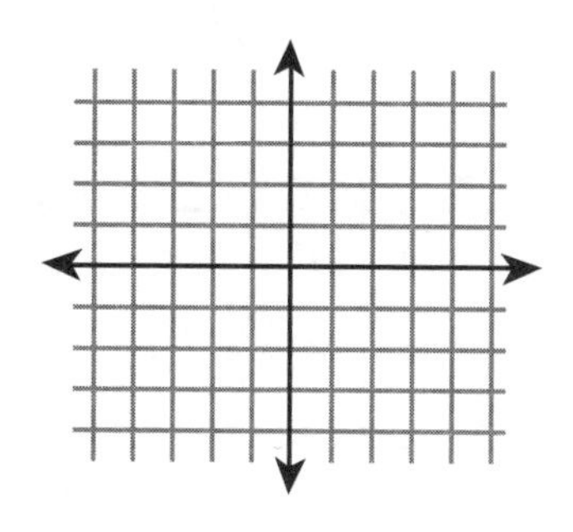

8. $y = x^2 + 8x + 20$

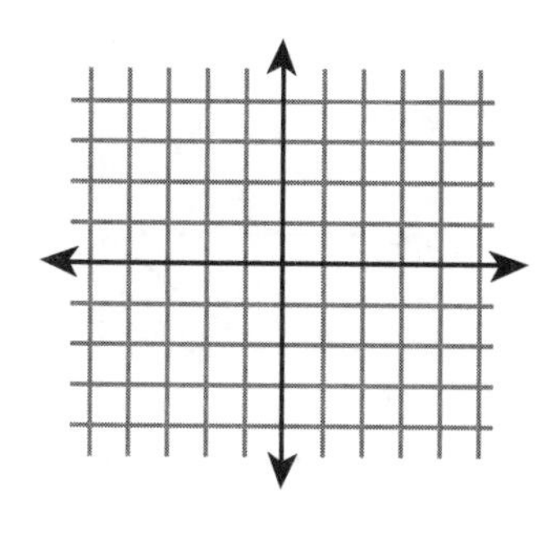

Name________________________________

HSN-VM.A.1

Describing Vectors

A **vector** is a quantity specified by magnitude and direction, such as velocity. A **scalar** is a quantity specified by magnitude only, such as speed. For example, 65 miles per hour due North is a vector known as velocity. 65 miles per hour is a scalar known as speed. When describing a vector, you must give both a magnitude and a direction.

Describe each vector.

Example:

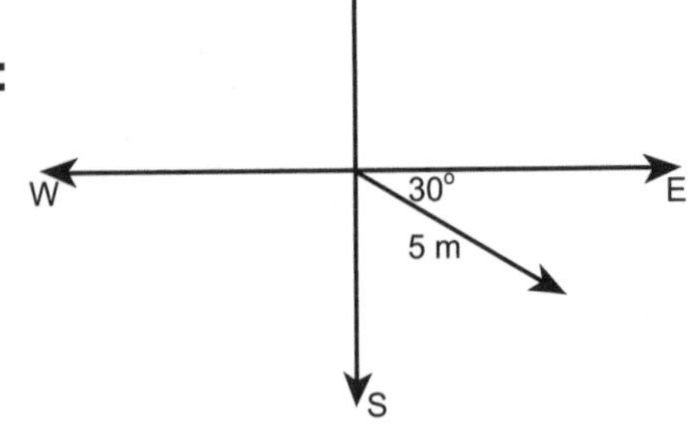

5 meters at 30 degrees south of east

5 meters at 30 degrees east of south

1.

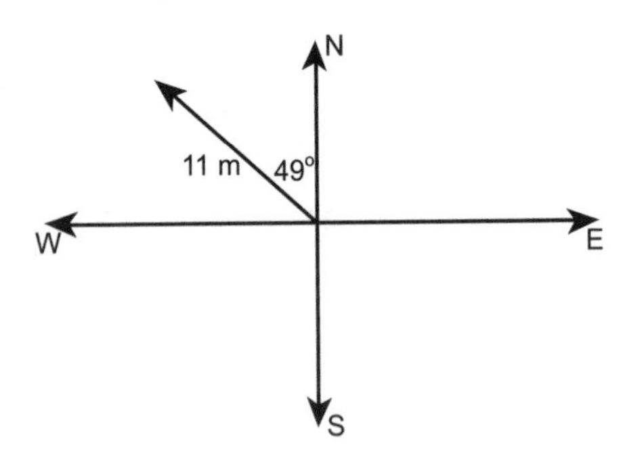

2.

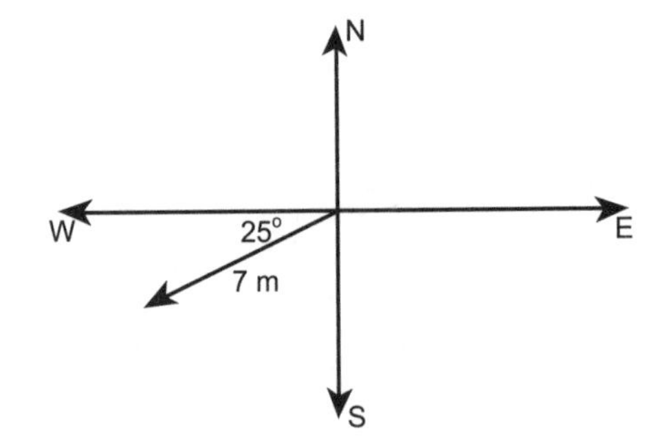

3.

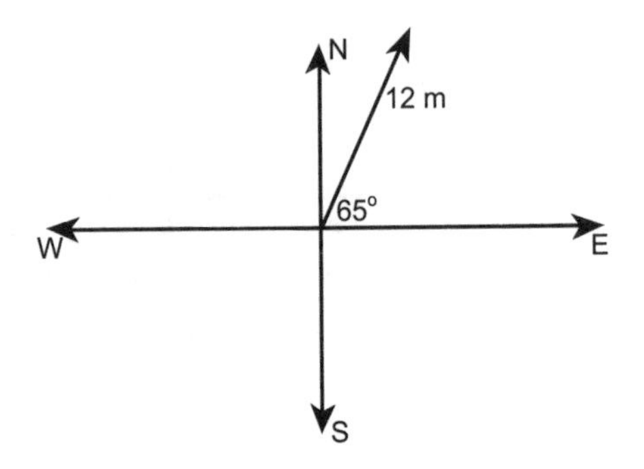

4.

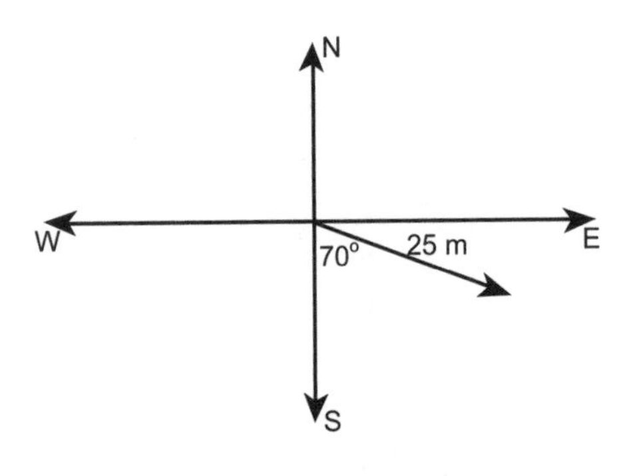

5.

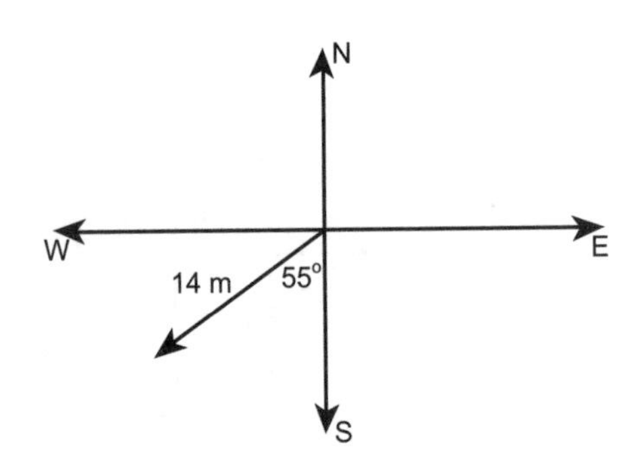

6.

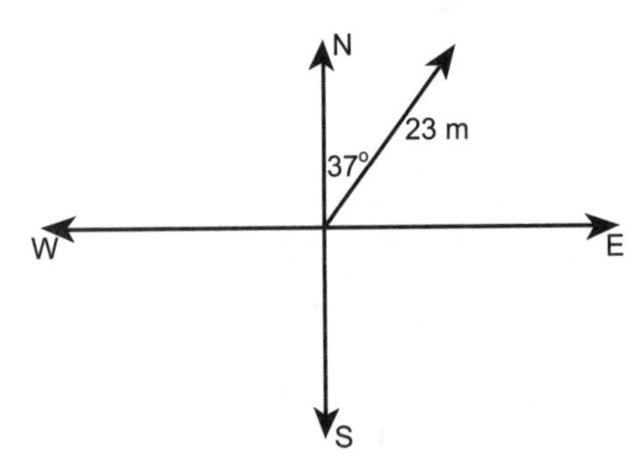

Adding Vectors Graphically

Vectors are commonly denoted as an arrow where the tip indicates the direction of the vector and the tail depicts the origin (see vector *u*).

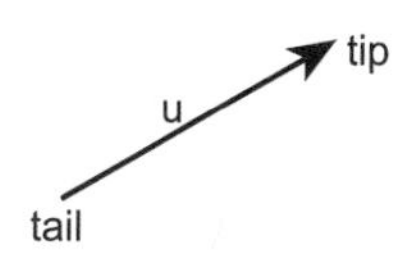

The length of the arrow is proportional to the magnitude of the vector. See examples *u* and *v* below.

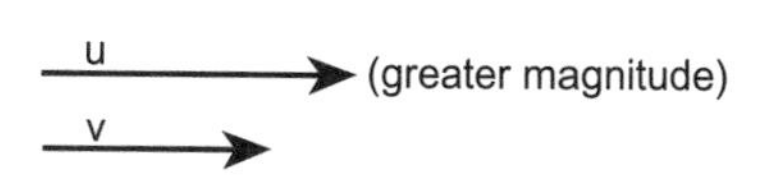

Using vectors *u*, *v*, *s*, and *t*, perform the following vector operations graphically.

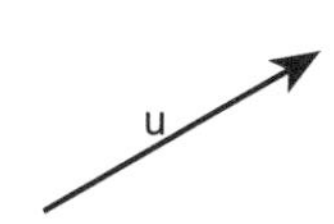

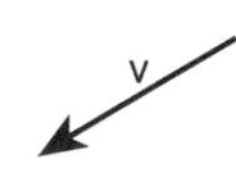

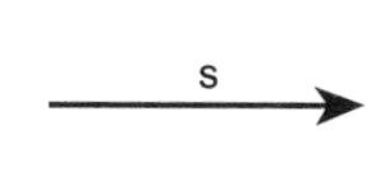

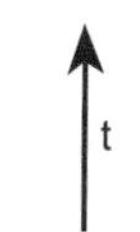

Example: $u + s$

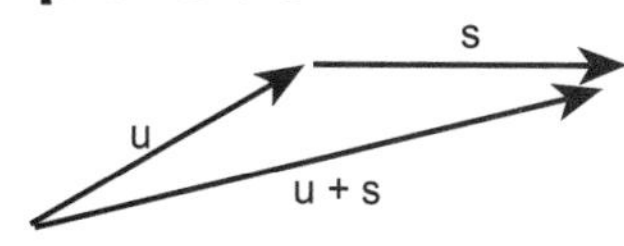

This method of addition is known as "tail to tip." Place vectors *u* and *s* together, then draw the resultant vector $u + s$ from the tail of *u* to the tip of *s*.

1. $u + t$

2. $v + s$

3. $s + t$

4. $u + v$

5. $s + s$

6. $v + u$

7. $t + u$

8. $s + v$

Is vector addition commutative? Explain.

Name______________________________ HSN-VM.B.4a, HSN-VM.B.5a

Vector Addition and Scalar Multiplication

Using the vectors, u, v, r, and t, perform the following vector operations.

Example: $2v - u$

a. Start with v and multiply its magnitude by two (make it twice as long). This is scalar multiplication.

b. Since a negative sign means the opposite, draw $-u$ in the exact opposite direction of vector u.

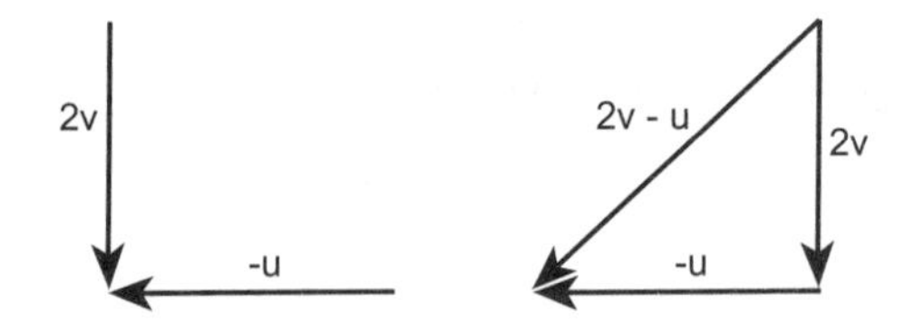

1. $r + \frac{1}{2}t$

2. $r - 2v$

3. $3v + \frac{1}{2}r$

4. $u - 2v$

5. $\frac{1}{2}t + r$

6. $5v + t$

7. $r + \frac{1}{2}t$

8. $r + 2v$

Is vector subtraction commutative? Explain.

Name________________________________

HSN.VM.A.2

Resolving Vectors

Resolve each vector into its *x* and *y* components. Round to the nearest tenth of a meter.

Example:

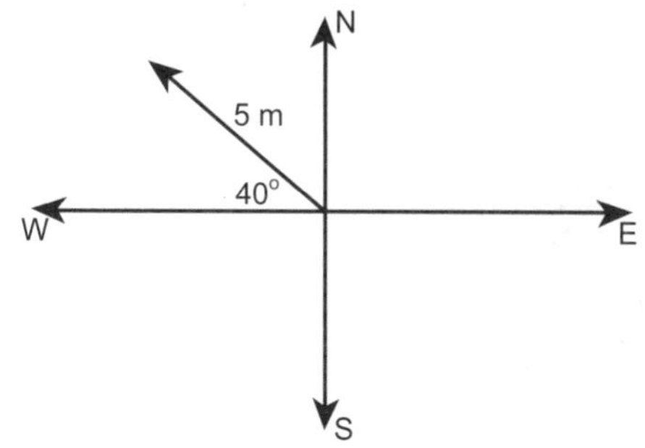

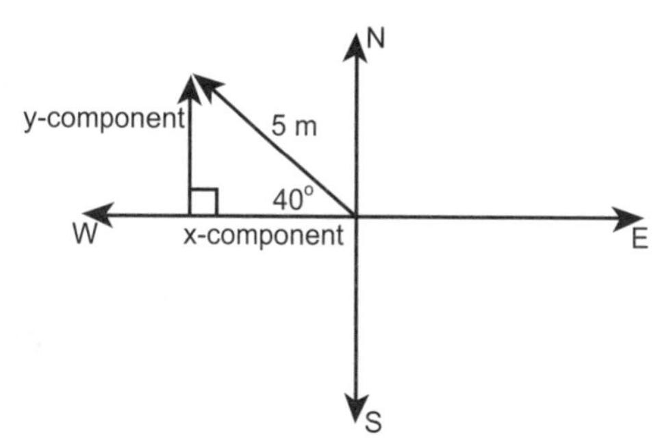

Note: The vector and its *x* and *y* components create a right triangle. Using the trigonometric ratios, solve for the *x*-component (*x*) and the *y*-component (*y*).

$\sin 40 = \frac{y}{5}$

$y = 5 \sin 40 = 3.213938 = 3.2 \text{ m}$

$\cos 40 = \frac{x}{5}$

$x = 5 \cos 40 = 3.830222 = 3.8 = {}^{-}3.8 \text{ m}$

Note: the *x*-component is negative since it points west.

1.

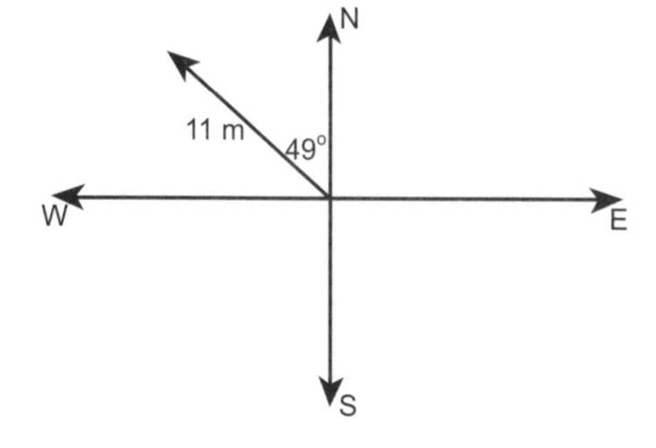

2.

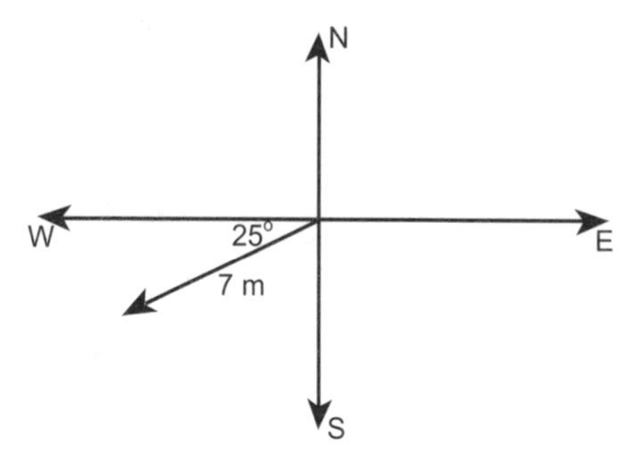

3.

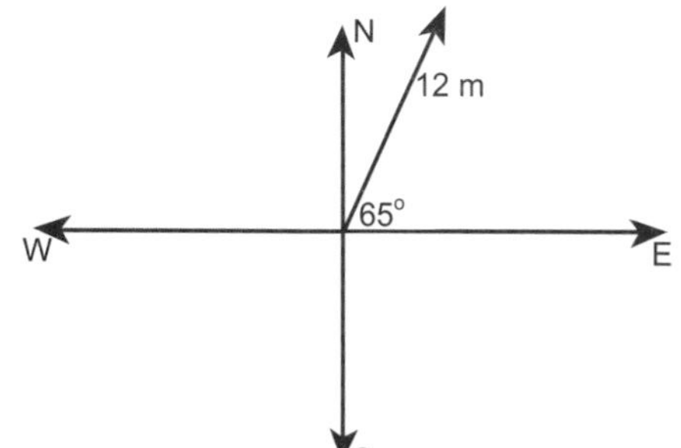

4.

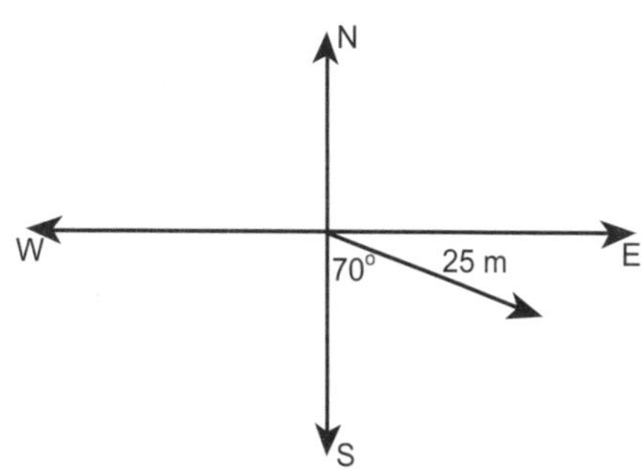

5.

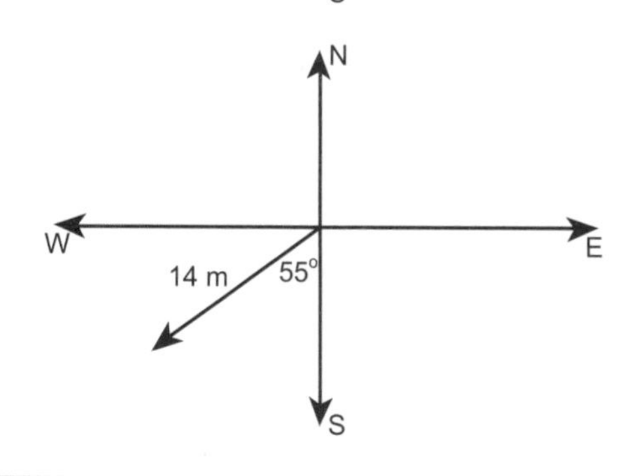

6.

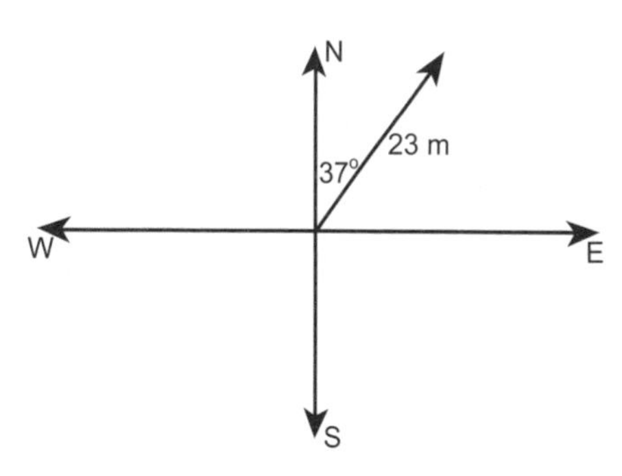

Name________________________________

Describing Vectors from *x* and *y* Components

Given the *x* and *y* components, describe the resultant vector. Round magnitudes to the nearest tenth and directions to the nearest degree.

Example:

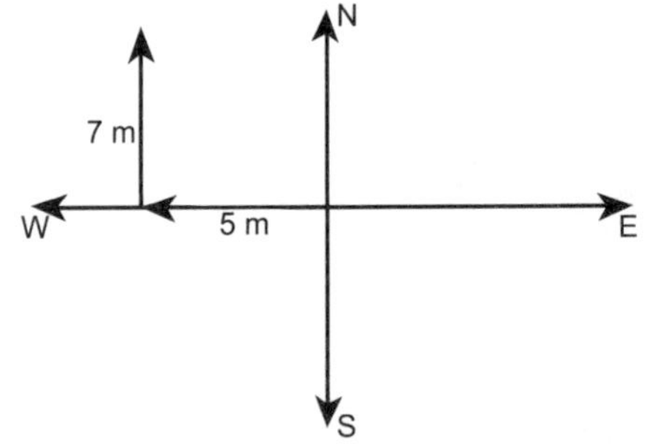

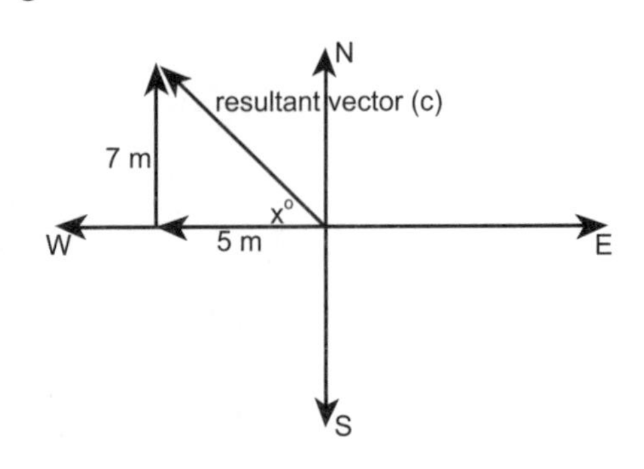

To find the magnitude, use the Pythagorean theorem

$c = \sqrt{a^2 + b^2} = \sqrt{7^2 + 5^2} = 8.602325 = 8.6\text{ m}$

To find the direction, use the trigonometric ratio.

$\tan x = \dfrac{\text{opposite}}{\text{adjacent}} = \dfrac{7}{5}$

$x = \tan^{-1}\left(\dfrac{7}{5}\right) = 54.4623222 = 54°$ North of West

1.

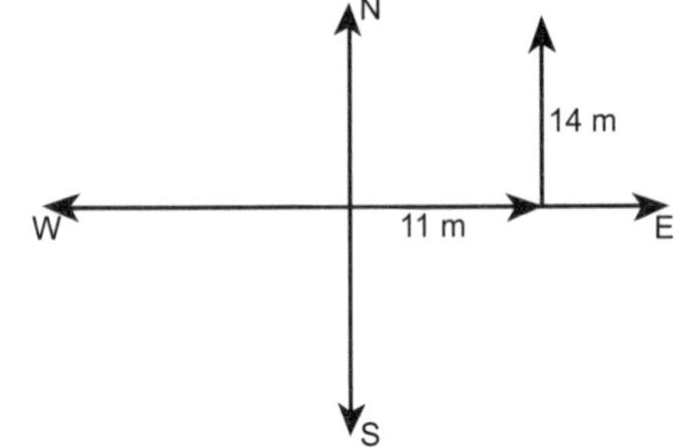

2.

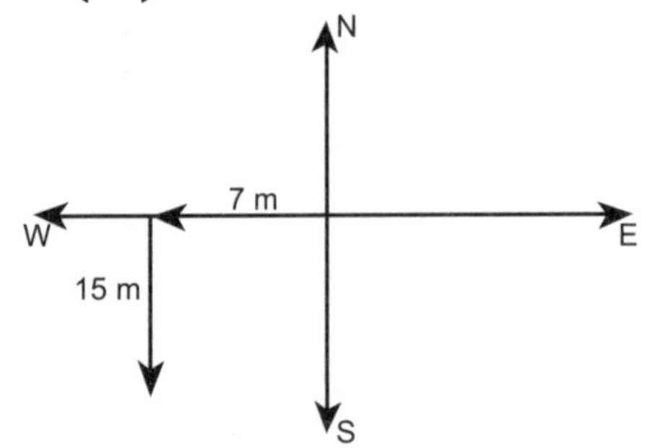

3.

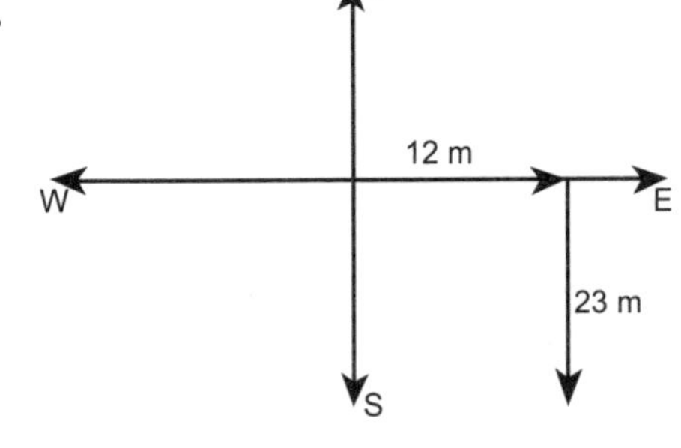

4.

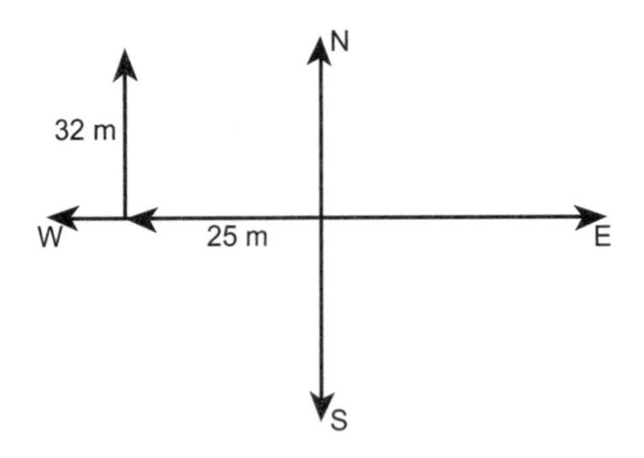

5.

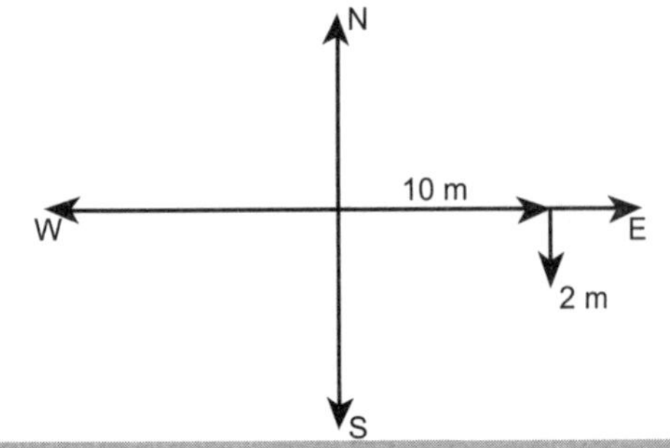

6.

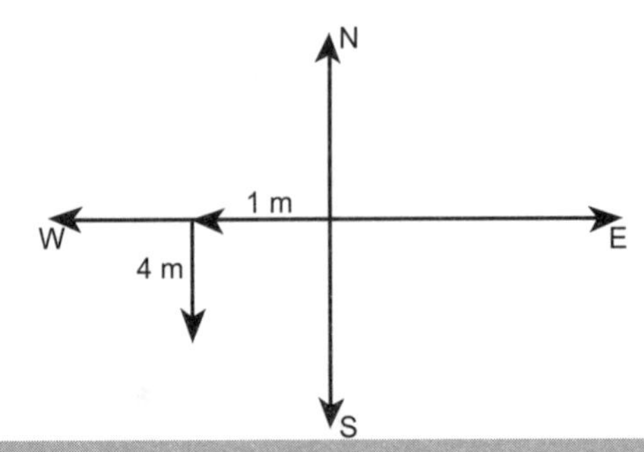

Name________________________________

HSN-VM.A.3

Problem Solving Involving Vectors

Draw a picture and solve each problem.

1. A boat is traveling directly across a river at 30 miles per hour. There is a current flowing down river at 5 miles per hour. What is the boat's resultant velocity vector?

2. An airplane pilot is traveling 500 miles per hour due south. He meets a wind current that is traveling 75 miles per hour due west. To continue flying due south, what adjustment in his navigation does he have to make?

3. A marathon runner runs 16 miles due north, then 10 miles due west. What is his displacement?

4. In an explosion, a piece of debris is tossed 99 meters northwest. What is the displacement in the north and west directions?

5. A duck flies 402 miles at 30 degrees southwest for the winter. How far south does the duck actually travel?

Name________________________________

HSN-VM.A.3

Adding Vectors Algebraically

Given two vectors, describe their resultant algebraically. Round the magnitude to the nearest tenth and direction to the nearest degree.

1. 15 m 45 degrees north of east and 5 m 10 degrees south of east.

2. 25 m 35 degrees south of east and 8 m 60 degrees north of east.

3. 99 m 30 degrees south of west and 15 m 53 degrees north of west.

4. 4 m 75 degrees north of east and 12 m 20 degrees north of west.

5. 62 m 23 degrees south of east and 36 m 45 degrees north of east.

Name________________________________

HSN-VM.A.3

Story Problems: Resolving Vectors

Draw a picture and solve each problem.

1. A glider is floating through the sky at 45 miles per hour with a heading of 50 degrees north of west when it meets a headwind of 10 miles per hour blowing 45 degrees south of east. What is its resultant velocity?

2. A riverboat is crossing the Mississippi River at 12 miles per hour due north. The current in the mighty Mississippi is 5 miles per hour at 43 degrees south of west. What is the resulting velocity of the riverboat?

3. A ship leaves port and sails 14 miles at 68 degrees south of west. Then, the ship turns due east and sails 8 miles. How far is the ship from port and what heading would be the shortest way home?

4. A pilot wants to fly 430 miles per hour at 45 degrees north of east, but there is a westerly wind blowing 20 miles per hour. What course correction does the pilot have to make to fly on his desired heading?

5. A long-distance swimmer starts out swimming a steady 2 miles per hour at 30 degrees south of west. A 5 mile per hour current is flowing at 10 degrees north of east. What is the swimmer's resultant velocity?

Answer Key

Page 7

1. $5x + 1$
2. $7a - b + 6$
3. $2a + 3b$
4. $2y^2 - 2ab$
5. $-2a^2 + 5ab - 14b^2$
6. $4x^2 + 3x + 1$
7. $6a^2 - 2ab - b^2$
8. $-12x^3 - 10x^2 + 5x - 4$
9. $6a + 8b - 11c$
10. 0
11. $-3x + 3y$
12. $5x^2 + y^2$
13. $-2x^2 + 3x - 2$
14. $x - 2y + 3$
15. $-5y^2 + 4y + 14$
16. $-2x + 15y - 2$
17. $6x - 6$
18. $6x - 8$

Page 8

1. $4y^2 + 4y - 14$
2. $3x^2 - 19x + 21$
3. $14x^2 - 12$
4. $13y^2 - 2$
5. $\frac{x^3}{2} + x^2 - x - 12$
6. $4x^2 - 12x + 9$
7. $21x^2 - 27x$
8. $49x^2 - 70x + 25$

Page 9

1. 13 of 25¢ and 15 of 29¢
2. 140 of \$3 and 105 of \$2
3. 3 quarters and 8 dimes
4. $2x^2 + 8xy - 2y^2$
5. $-4ab - 5ac + 7bc$
6. $6x^2 + 6y^2 - 3z^2$
7. P = 10c + 8 units
8. square

Page 10

1. $x = 10$
2. $y = 12$
3. $n = 3$
4. $b = \frac{1}{2}$
5. $x = 8$
6. $y = 2$
7. $x = 1$
8. $r = -3$
9. $y = 15$
10. $x = 3$
11. $x = -2$
12. $t = 3$
13. $w = 20$
14. $x = -9$

Page 11

1. Correction: $16c^2$
2. No errors.
3. No errors.
4. No errors.
5. No errors.
6. Correction: $16b^2c^2$
7. No errors.
8. Correction: $-72x^5y$
9. Correction: $8a^7$
10. Correction: $12s^3t^2$
11. No errors.
12. Correction: $-9x^3z^3$
13. No errors.
14. No errors.
15. Correction: $2x^6y^4z^4$
16. Correction: $4u^8v^3w$
17. No errors.
18. No errors.
19. Correction: $-36x^5y^3$
20. No errors.
    Explanation: Take the power of each factor. Remember that when you calculate the power of a power, you multiply exponents: $(-5x^2)^3 = (-5)^3 (x^2)^3 = -125x^6$.

Page 12

1. cubic binomial
2. fifth degree binomial
3. tenth degree monomial
4. sixth degree trinomial
5. sixth degree binomial
6. 15th degree monomial
7. cubic trinomial
8. fourth degree binomial
9. linear monomial
10. seventh degree binomial
11. sixth degree binomial
12. cubic trinomial
13. fourth degree trinomial
14. linear binomial

Page 13

1. $7x^4 + x^3 + 0x^2 + x + 0$
2. $6x^3 + x^2 + 0x + 23$
3. $19x^6 + 0x^5 + 0x^4 + 0x^3 + 0x^2 + 0x + 1$
4. $14x^2 + 4x + 0$
5. $x^3 + 0x^2 + 0x + 0$
6. $3x^2 + 10x + 0$
7. $x^2 + 2x + 1$
8. $8x^3 + 0x^2 + 4x + 0$
9. $x + 1$
10. $3x^3 + 2x^2 + x + 0$
11. $x^4 + 0x^3 + 0x^2 + 20x + 18$
12. $25x^2 + 5x + 0$
13. $18x^3 + 0x^2 + x + 0$

Answer Key

Page 14

1. = $^{-}231$ **2.** = $^{-}314$ **3.** = 14 **4.** = 65 **5.** = $^{-}62$ **6.** = 8 **7.** = 401 **8.** = 1082 **9.** = 785 **10.** = $^{-}18$

Cross out the correct answers below. Use the remaining letters to complete a statement, then rewrite the statement as a common adage.

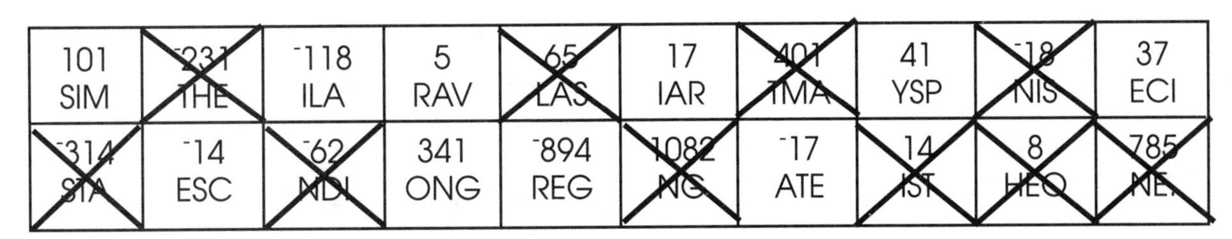

101 SIM	$^{-}231$ THE	$^{-}118$ ILA	5 RAV	65 LAS	17 IAR	401 TMA	41 YSP	$^{-}18$ NIS	37 ECI
$^{-}314$ STA	$^{-}14$ ESC	$^{-}62$ NDI	341 ONG	$^{-}894$ REG	1082 NG	$^{-}17$ ATE	14 IST	8 HEO	785 NE

S I M I L A R A V I A R Y S P E C I E S C O N G R E G A T E.

Common adage: Birds of a feather flock together.

Page 15

1. $2x^2 - 2xy + 6y^2$
2. $-8n - 10n^4$
3. $c^4d^4 + 2c^3d^3 + c^2d^2$
4. $4xy^2 - 2x^2y^2 - 2x^3y^3$
5. $-2a^3b + 6a^2b^2 + 4ab^3$
6. $24n^3 - 6n^2$
7. $-w^2z^3 + 2wz^3 - z^3$
8. $-3a^4b^4 + 6a^3b^3$
9. $36x^4y - 24x^3y^3 - 28x^2y$
10. $-12k^3m^2 + 18k^2m^3 - 24k^3m^3 + 6k^4m^4$
11. $-4n^4 - n^3$
12. $-4x^3 + 7x^2$
13. $2x^5 - 4x^4 + 16x^3 - 10x^2$
14. $-18x^5 + 6x^3$
15. $15x^3 - 18x^2 - 24x$
16. $-10x^5 - 15x^4 + 35x^3 - 45x^2$
17. $a = 2x^2 + 8x$
18. $3xy + 6y$

Page 16

1. $z^2 - 9$
2. $3t^2 - 11t + 6$
3. $a^2 + 10a + 25$
4. $2ax + 2bx + ay + by$
5. $x^2 - \frac{3}{2}xy - y^2$
6. $16x^2 - 25$
7. $0.32n^2 - 9.8n + 45$
8. $2c^3 + 4c^2 + 6cd + c^2d + 2d^2$
9. $9a^4 - 4b^4$
10. $h^3 - h^2k + hk^2 + 3k^3$
11. $2x^3 + x^2 + 5x - 3$
12. $x^4 + 2x^3 - x^2 - 3x + 1$
13. $-m^3 + m^2n - mn^2 + n^3$
14. $y^3 - y^2 + 2$
15. $\frac{1}{6}x^2 + x - 12$
16. $3x^3 + 11x^2 - 27x - 35$
17. $2x^4 + 3x^3 - x^2 - 9x - 15$
18. $12x^3 - 10x^2 + 8$

Page 17

1. $11x^2 + 5x + 6$
2. $y^2 + 11y + 28$
3. $x^2 - 4x - 32$
4. $x^2 - 12x + 32$
5. $y^2 + y - 20$
6. $x^2 - 11x + 18$
7. $2x^2 + 10x + 12$
8. $6x^2 + 19x + 10$
9. $12x^2 - 23x - 9$
10. $8x^2 + 14x - 15$
11. $3n^2 - 23n + 14$
12. $15x^2 - 29x - 14$
13. $8x^2 + 2x - 15$
14. $-3x^2 - 16x - 16$
15. $2x^2 + 7xy + 6y^2$
16. $18x^2 - 15xy + 2y^2$
17. $12x^2 - 13xy - 4y^2$
18. $20a^2 + 7ab - 3b^2$

Page 18

1. O
2. R
3. T
4. M
5. I
6. N
7. I
8. L
9. S
10. A

All but one of the answers are: T R I N O M I A L S.
Explanation: It is the sum and difference of the same two terms.

Answer Key

Page 19

1. $x^2 - 9$
2. $y^2 - 100$
3. $a^2 - 16$
4. $x^2 - 49$
5. $4x^2 - 1$
6. $25y^2 - 36$
7. $16x^2 - 9$
8. $9n^2 - 49$
9. $9c^2 - 16$
10. $4x^2 - 81$
11. $49x^2 - 25$
12. $x^2 - y^2$
13. $25x^2 - y^2$
14. $4x^2 - 121y^2$
15. $9x^2 - 49y^2$
16. $4x^2 - y^4$
17. $9x^4 - 1$
18. $x^4 - y^4$
19. difference
20. The products of the outside and inside terms are opposites and cancel each other out (equal 0).

Page 20

1. $x^2 - 16x + 64$
2. $a^2 + 10a + 25$
3. $x^2 - 6x + 9$
4. $9n^2 + 6n + 1$
5. $y^2 - 20y + 100$
6. $9x^2 + 12x + 4$
7. $16x^2 - 24x + 9$
8. $4a^2 + 20a + 25$
9. $36x^2 + 12x + 1$
10. $25b^2 + 20b + 4$
11. $16x^2 - 8xy + y^2$
12. $36x^2 - 60xy + 25y^2$
13. $9y^2 - 30yz + 25z^2$
14. $49a^2 + 28ab + 4b^2$
15. $121x^2 - 44xy + 4y^2$
16. $25a^2 + 30ab + 9b^2$

Page 21

1.
$$\begin{array}{r} 2x^4 - 3x^3 \qquad \\ - 2x^3 + 3x^2 \\ \hline 2x^4 - 5x^3 + 3x^2 \end{array}$$

2. $= (x+1)(x+1)(x+1)$
$= (x+1)(x^2+2x+1)$
$$\begin{array}{r} x^3 + 2x^2 + x \qquad \\ + x^2 + 2x + 1 \\ \hline x^3 + 3x^2 + 3x + 1 \end{array}$$

3.
$$\begin{array}{r} x^3 + 6x^2 + 10x \qquad \\ + x^2 + 6x + 10 \\ \hline x^3 + 7x^2 + 16x + 10 \end{array}$$

4.
$$\begin{array}{r} 2x^4 - 12x \qquad \\ + x^3 - 6 \\ \hline 2x^4 + x^3 - 12x - 6 \end{array}$$

5. $x^4 - 1$

6.
$$\begin{array}{r} 54x^3 - 9x^2 + 9x \qquad \\ - 24x^2 + 4x - 4 \\ \hline 54x^3 - 33x^2 + 13x - 4 \end{array}$$

7.
$$\begin{array}{r} 5x^3 + x^2 - 8x \qquad \\ - 5x^2 - x + 8 \\ \hline 5x^3 - 4x^2 - 9x + 8 \end{array}$$

8.
$$\begin{array}{r} 6x^3 + 2x^2 + x \qquad \\ -24x^2 - 8x - 4 \\ \hline 6x^3 - 22x^2 - 7x - 4 \end{array}$$

9. $= (2x - 3)(2x - 3)(2x - 3)$
$= (2x - 3)(4x^2 - 12x + 9)$
$$\begin{array}{r} 8x^3 - 24x^2 + 18x \qquad \\ - 12x^2 + 36x - 27 \\ \hline 8x^3 - 36x^2 + 54x - 27 \end{array}$$

10. $= (3x^2 + 1)(3x^2 + 1)(3x^2 + 1)$
$= (3x^2 + 1)(9x^4 + 6x^2 + 1)$
$$\begin{array}{r} 27x^6 + 18x^4 + 3x^2 \qquad \\ + 9x^4 + 6x^2 + 1 \\ \hline 27x^6 + 27x^4 + 9x^2 + 1 \end{array}$$

11.
$$\begin{array}{r} 14x^4 + 14x^4 - 98x \qquad \\ + x^3 + x^2 - 7 \\ \hline 14x^4 + 15x^3 + x^2 - 98x - 7 \end{array}$$

12.
$$\begin{array}{r} {}^{-}6x^4 - 4x^2 - 2x \qquad \\ 3x^3 + 2x + 1 \\ \hline {}^{-}6x^4 + 3x^3 - 4x^2 + 1 \end{array}$$

13. $(11x^2 - 1)(11x^2 - 1)(11x^2 - 1)$
$(11x^2 - 1)(121x^4 - 22x^2 + 1)$
$$\begin{array}{r} 1331x^6 - 242x^4 + 11x^2 \qquad \\ - 121x^4 + 22x^2 - 1 \\ \hline 1331x^6 - 363x^4 + 33x^2 - 1 \end{array}$$

14. $({}^{-}2x^2 + x)(4x^4 - 4x^3 + x^2)$
$$\begin{array}{r} {}^{-}8x^6 + 8x^5 - 2x^4 \qquad \\ 4x^5 - 4x^4 + x^3 \\ \hline {}^{-}8x^6 + 12x^5 - 6x^4 + x^3 \end{array}$$

Answer Key

Page 22

1. $(2x + 1)(2x - 1)$
2. $(x - 3)(x + 3)$
3. $9(4x^2 - 1)$
   $9(2x - 1)(2x + 1)$
4. $(10x - 9)(10x + 9)$
5. $(5x - 2)(5x + 2)$
6. $(9x - 11)(9x + 11)$
7. $(x - 4)(x + 4)$
8. $16(9x^2 - 1)$
   $16(3x - 1)(3x + 1)$
9. $(x - 5)(x + 5)$
10. $(25 - 4x)(25 + 4x)$
11. $(10 - x)(10 + x)$
12. $(x - 6)(x + 6)$
13. $(11x - 7)(11x + 7)$
14. $(7x - 4)(7x + 4)$

$(x + 13)(x - 13)$ THE	~~$16(3x - 1)(3x + 1)$ SUM~~	~~$(x - 4)(x + 4)$ OFA~~	$(6x + 5)(6x - 5)$ PRO	~~$(25 - 4x)(25 + 4x)$ QUO~~	$(x + 1)(x - 1)$ DUC
$(9 + x)(9 - x)$ TOF	~~$9(2x - 1)(2x + 1)$ TIE~~	$(x + 7)(x - 7)$ THE	~~$(2x + 1)(2x - 1)$ NTA~~	$(9x + 1)(9x - 1)$ SUM	$(x + 2)(x - 2)$ AND
~~$(10 - x)(10 + x)$ WAS~~	$(5x + 3)(5x - 3)$ DIF	~~$(x - 5)(x + 5)$ HAS~~	$(8x + 1)(8x - 1)$ FER	~~$(11x - 7)(11x + 7)$ MAN~~	~~$(x - 6)(x + 6)$ NER~~
$(x + 18)(x - 18)$ ENC	~~$(10x - 9)(10x + 9)$ THA~~	~~$(x - 3)(x + 3)$ TIS~~	~~$(5x - 2)(5x + 2)$ MYP~~	$(7x + 11)(7x - 11)$ EOF	$(x + 8)(x - 8)$ THE
$(x + 15)(x - 15)$ SQU	~~$(9x - 11)(9x + 11)$ ROB~~	$(x + 9)(x - 9)$ ARE	$(3x + 2)(3x - 2)$ ROO	~~$(7x - 4)(7x + 4)$ LEM~~	$(x + 9)(x - 9)$ TS.

15. The factored form of the difference of the two squares is

T H E P R O D U C T O F T H E S U M A N D D I F F E R E N C E O F T H E S Q U A R E R O O T S .

Page 23

1. $(x + 4)^2$
2. $(x - 8)^2$
3. $(y + 6)^2$
4. $(y - 5)^2$
5. $(4x + 1)^2$
6. $(3x - 1)^2$
7. $(5x + 1)^2$
8. $(9y - 5)^2$
9. $(2y - 5)^2$
10. $(5x + 6)^2$
11. $(4 + 5y)^2$
12. $(4y + 3)^2$
13. $(7x - 1)^2$
14. $(3y - 5)^2$
15. $(y + 2)^2$
16. $81(y + a)^2$
17. $4, (7y - 2)^2$
18. $36y, (3y + 6)^2$
19. $22y, (y - 11)^2$
20. $4, (2y + 5)^2$

Explanation: Multiplying 2 positives or 2 negatives always equals a positive number.

Page 24

1. two numbers: 2, 4
2. two numbers: 2, 3
3. two numbers: -2, -7 or $(x - 2)(x - 7)$
4. two numbers: -2, 18 or $(x - 2)(x + 18)$
5. two numbers: -3, -5 or $(x - 3)(x - 5)$
6. two numbers: -8, 4 or $(x - 8)(x + 4)$
7. two numbers: -3, 2 or $(x - 3)(x + 2)$
8. two numbers: -3, 6 or $(x - 3)(x + 6)$
9. two numbers: -2, 9 or $(x - 2)(x + 9)$
10. two numbers: -7, 8 or $(x - 7)(x + 8)$
11. two numbers: -25, 3 or $(x - 25)(x + 3)$
12. two numbers: -8, 5 or $(x - 8)(x + 5)$

Page 25

1. $(x + 4)(x + 3)$
2. $(m + 3)(m + 7)$
3. $(y - 8)(y + 1)$
4. $(x - 1)(x - 5)$
5. $(x + 8)(x - 4)$
6. $(x - 5)(x + 3)$
7. $(x - 2)(x - 4)$
8. $(y + 3)(y + 6)$
9. $(t - 1)(t - 3)$
10. $(v + 2)(v + 10)$
11. $(k - 3)(k - 17)$
12. $(a - 2b)(a - 12b)$
13. $(y - 6)(y + 12)$
14. $(x - 15y)(x + 4y)$
15. $(5r - s)(3r + s)$
16. $3(x - 2y)(x + 9y)$
17. $(x - 6y)(x + 1y)$
18. $(x + 6y)(x + 2y)$
19. $(y - 5x)(y - 2x)$
20. $(a - 15b)(a + 4b)$

Answer Key

Page 26

1. $(2x + 1)(x - 3)$
2. $(3x - 2)(x + 4)$
3. $(2y + 1)(y + 7)$
4. $(7a - 4)(a - 1)$
5. $(5n + 2)(n + 3)$
6. $(2y + 3)(2y + 1)$
7. $(3x + 7)(x - 1)$
8. $(2x + 3)(x + 5)$
9. $(3y - 2)(3y + 4)$
10. $(3x + 4)(2x - 5)$
11. $(2n - 7)(n + 2)$
12. prime
13. $(2x + 5)(5x - 6)$
14. $(3y + 1)(4y + 1)$
15. $(2n - 1)(n + 5)$
16. $(2x + 3)(x + 2)$
17. $(5a + 3)(a - 9)$
18. $(3x - 8)(5x + 4)$
19. $(2a - 1)(4a - 3)$
20. $(2y + 5)(y - 4)$

Page 27

1. $(x - 6)(x - 6)$
2. $(x + 12)(x + 12)$
3. $(x - 18)(x + 2)$
4. $(x - 11)(x + 2)$
5. $(x + 16)(x + 2)$
6. $(x - 8)(x + 7)$
7. $(3x + 2)(2x + 1)$
8. $(3x + 8)(x - 2)$
9. $(3x - 4)(2x + 1)$
10. $(5x - 2)(3x + 1)$
11. $(6x + 1)(3x + 1)$
12. $(5x + 2)(4x + 1)$
13. $(5x - 1)(x - 5)$
14. $(x - 10)(x + 1)$

Page 28

1. $x(x^2 + 5x + 6)$
   $= x(x + 2)(x + 3)$
2. $a = x \quad b = 1$
   $(x + 1)(x^2 - x + 1)$
3. $a = 4x \quad b = 3$
   $(4x - 3)(16x^2 + 12x + 9)$
4. $2x(x^2 - 4x + 4)$
   $2x(x - 2)(x - 2)$
5. $(x - 4)(x + 2)$
6. $a = x \quad b = 6$
   $x(x^3 + 216)$
   $x(x + 6)(x^2 - 6x + 36)$
7. $(x + 1)(10x - 5)$
8. $x(2x^2 - 7x + 6)$
   $x(2x - 3)(x - 2)$
9. $2x(125x^3 - 27)$
   $2x(5x - 3)(25x^2 + 15x + 9)$
10. $10x^2(4x^2 - 1)$
    $10x^2(2x - 1)(2x + 1)$

Page 29

1. m^5
2. $\frac{x}{2}$
3. $\frac{2b}{a}$
4. $\frac{3}{-2u^2v^2}$
5. $\frac{a^9}{-2}$
6. -1
7. $-4x^2z$
8. $\frac{1}{2x^2y^3}$
9. x^4
10. $\frac{3x}{4}$
11. $14t^2$
12. $\frac{4c^3}{a^2b^2}$
13. $\frac{-5x}{3y}$
14. $\frac{2}{3n}$
15. $\frac{c^4d}{3}$
16. $\frac{6z^2}{-11x^2}$

Page 30

1. $a + 2$
2. $2x + 5$
3. $2y + 3$
4. $x - y$
5. $-5u^2 + 3u + 1$
6. $4x - 3x^2 + 2x^3$
7. $mn + \frac{1}{n} - \frac{1}{m}$
8. $-38^3 + 4ab + 1$
9. $7k^8m - 2k + 6m$
10. $\frac{4v^2}{u} - \frac{9v}{u} + 6$
11. $x - 5y$
12. $x^2y - 2xy + 2y$
13. $2z - 1 + \frac{1}{z}$
14. $\frac{1}{a} + \frac{7}{a^2} + \frac{12}{a^3}$
15. $x - 1$
16. $3mn^3 - 2n^2 + \frac{4n}{m^2}$

Page 31

1. $s - 1$
2. $a - 1 + \frac{6}{a + 13}$
3. $x - 2 + \frac{8}{x + 2}$
4. $c + 1 + \frac{c + 2}{3c + 2}$
5. $3r + 5 + \frac{10}{2r - 3}$
6. $3t - 2 + \frac{5}{3t + 2}$
7. $2u + 3v$
8. $z^2 - 2z + 3$
9. $3x^2 - 2x + 3$
10. $2y + 3$
11. $x^2 - 3x + 4 + \frac{2}{x + 2}$
12. $2x^2 + x + 1$
13. $y^2 - 3y + 3$
14. $x^2 + 3x - 7$

Answer Key

Page 32

1.

$$\begin{array}{r} x \\ x^2 - 1\overline{)x^3 - 1} \\ -x^3 + x \\ x - 1 \end{array}$$

$$= x + \frac{x - 1}{x^2 - 1}$$

2.

$$\begin{array}{r} 2x + 1 \\ x - 3\overline{)2x^2 - 5x - 3} \\ -2x^2 + 6x \\ x - 3 \\ x - 3 \\ 0 \end{array}$$

3.

$$\begin{array}{r} x - 5 \\ x + 2\overline{)x^2 - 3x - 7} \\ -x^2 + 2x \\ -5x - 7 \\ +5x + 10 \\ 3 \end{array}$$

$$= x - 5 + \frac{3}{x + 2}$$

4.

$$\begin{array}{r} x^3 + x + 1 \\ x - 1\overline{)x^2 - 6} \\ -x^3 + x^2 \\ x^2 - 6 \\ -x^2 + x \\ x - 6 \\ -x + 1 \\ -5 \end{array}$$

$$= x^2 + x + 1 + \frac{^-5}{x - 1}$$

5.

$$\begin{array}{r} x^2 - 8x + 16 \\ x + 2\overline{)x^3 - 6x^2 + 1} \\ -x^3 - 2x^2 + 1 \\ -8x^2 + 1 \\ +8x^2 + 16x \\ 16x + 1 \\ -16x + 32 \\ -31 \end{array}$$

$$= x^2 - 8x + 16 + \frac{^-31}{x + 2}$$

6.

$$\begin{array}{r} 5x + 1 \\ x - 7\overline{)5x^2 - 34x - 7} \\ -5x^2 + 35x \\ x - 7 \\ x - 7 \\ 0 \end{array}$$

7.

$$\begin{array}{r} x^3 - 5x^2 + 10x - 25 \\ x + 2\overline{)x^4 - 3x^3 - 5x - 6} \\ -x^4 - 2x^3 \\ -5x^3 - 5x \\ +5x^3 + 10x^2 \\ +10x^2 - 5x \\ -10x^2 - 20x \\ -25x - 6 \\ 25x + 50 \\ 44 \end{array}$$

$$= x^3 - 5x^2 + 10x - 25 + \frac{44}{x + 2}$$

8.

$$\begin{array}{r} 2x - 1 \\ 3x + 1\overline{)6x^2 - x - 7} \\ -6x^2 - 2x \\ -3x - 7 \\ +3x + 1 \\ -6 \end{array}$$

$$= 2x - 1 + \frac{^-6}{3x + 1}$$

Page 34

1. $2x^3 - 7x^2 - 8x + 16$

$$\begin{array}{r|rrr|r} 4 & 2 & -7 & -8 & 16 \\ & & 8 & 4 & -16 \\ \hline & 2 & 1 & -4 & 0 \end{array}$$

$= 2x^2 + x - 4$

2. $2x^2 + 5x + 3$

$$\begin{array}{r|rr|r} -1 & 2 & 5 & 3 \\ & & -2 & -3 \\ \hline & 2 & 3 & 0 \end{array}$$

$= 2x + 3$

3. $x^2 + 3x - 18$

$$\begin{array}{r|rr|r} -3 & 1 & 3 & -18 \\ & & -3 & 0 \\ \hline & 1 & 0 & -18 \end{array}$$

$$= x + \frac{^-18}{x + 1}$$

4. $2x^2 + 4x + 3$

$$\begin{array}{r|rr|r} 3 & 2 & -4 & 3 \\ & & 6 & 6 \\ \hline & 2 & 2 & 9 \end{array}$$

$$= 2x + 2 + \frac{9}{x - 3}$$

5. $x^3 - x^2 + x + 8$

$$\begin{array}{r|rrr|r} 1 & 1 & -1 & 1 & 8 \\ & & 1 & 0 & 1 \\ \hline & 1 & 0 & 1 & 9 \end{array}$$

$$= x^2 + 1 + \frac{9}{x - 1}$$

6. $3x^3 - x^2 + 2x - 4$

$$\begin{array}{r|rrr|r} 2 & 3 & -1 & 2 & -4 \\ & & 6 & 10 & 24 \\ \hline & 3 & 5 & 12 & 20 \end{array}$$

$$= 3x^2 + 5x + 12 + \frac{20}{x - 2}$$

Page 36

1. $(x - 2)(2x - 1)(x + 1)$

2. $(x - 1)(x + 1)(x - 1)$

3. $(x - 1)(x + 5)(x + 3)$

4. $(x + 2)(x - 4)(x + 1)$

5. $(x + 1)(2x - 3)(x - 1)$

6. $(x - 5)(x + 1)(3x - 1)$

Answer Key

Page 38

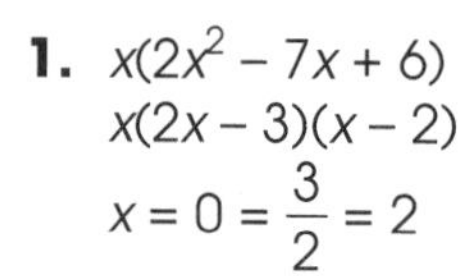

1. $x(2x^2 - 7x + 6)$
   $x(2x - 3)(x - 2)$
   $x = 0 = \frac{3}{2} = 2$

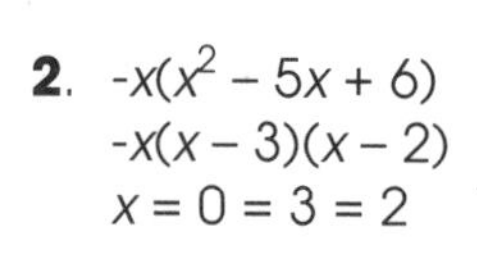

2. $-x(x^2 - 5x + 6)$
   $-x(x - 3)(x - 2)$
   $x = 0 = 3 = 2$

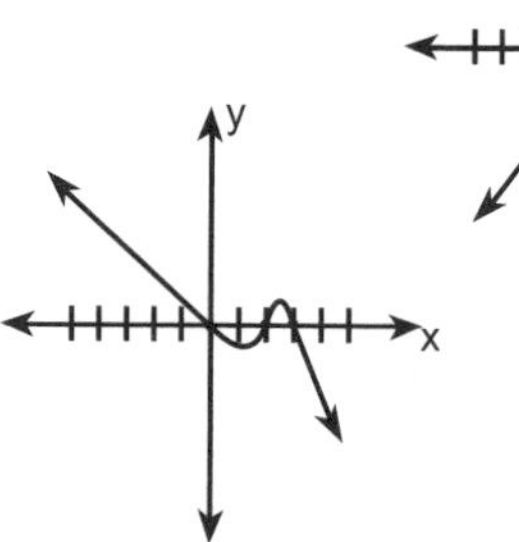

3. $^-4x^2(x + 3) + 1(x + 3)$
   $(1 + 2x)(1 - 2x)(x + 3)$
   $x = {^-3} = \pm \frac{1}{2} = 0.5$

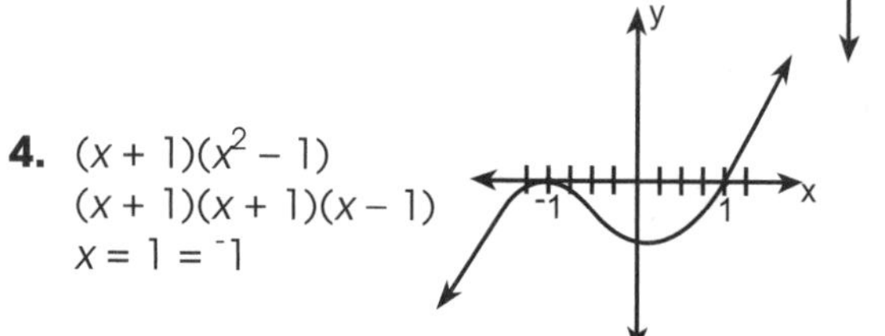

4. $(x + 1)(x^2 - 1)$
   $(x + 1)(x + 1)(x - 1)$
   $x = 1 = {^-1}$

5. $10x^2(4x^2 - 1)$
   $10x^2(2x - 1)(2x + 1)$
   $x = 0 = \frac{^-1}{2} = \frac{1}{2}$

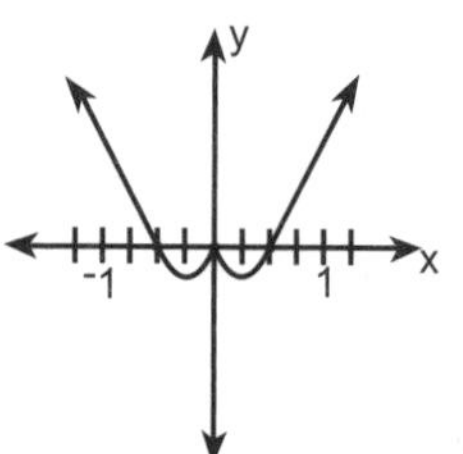

6. $x^2(x^2 - 9x + 20)$
   $x^2(x - 5)(x - 4)$
   $x = 0 = 5 = 4$

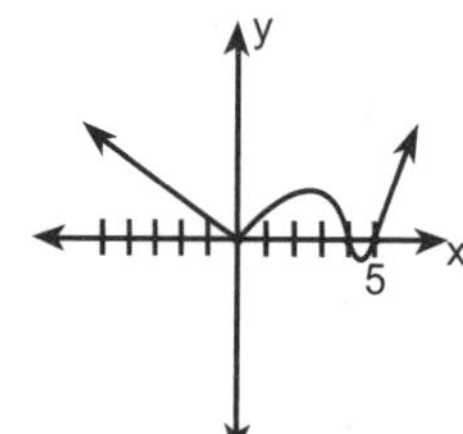

Page 39

1. $x = 343$
2. $x = 10$ or $^-10$
3. $x = 2$
4. $x = \pm 5$
5. $x = 125$
6. $x = 25$
7. $x = 10{,}000$
8. $x = 3$
9. $x = {^-19{,}683}$
10. $x = {^-4}$
11. $x = 43{,}046{,}721$

Non-real roots of an equation are the

I	M	A	G	I	N	A	R	Y		R	O	O	T	S
2	3	4	7	2	1	4	5	8		5	6	6	9	10

Page 40

1. $(2x + 6)^5 = {^-32}$
   $2x + 6 = {^-2}$
   $2x = {^-8}$
   $x = {^-4}$
2. $(2x + 1)^{\frac{1}{3}} = {^-7}$
   $2x + 1 = {^-343}$
   $2x = {^-344}$
   $x = {^-172}$
3. $(x - 1)^2 = 144$
   $x - 1 = 12$ or $^-12$
   $x = 13$ or $^-11$
4. $(x - 2)^2 = 16$
   $x - 2 = \pm 4$
   $x = 6$ or $^-2$
5. $(2x - 5)^4 = 6561$
   $2x - 5 = \pm 9$
   $x = 7$ or $^-2$
6. $5x - 1 = 7$ or $^-7$
   $x = \frac{8}{5}$ or $\frac{^-6}{5}$
7. $(10x - 18)^{\frac{1}{2}} = 49$
   $10x - 18 = 2401$
   $10x = 2419$
   $x = 241.9$
8. $(12x - 1)^3 = 216$
   $12x - 1 = 6$
   $12x = 7$
   $x = \frac{7}{12}$
9. $(4x + 3)^{\frac{1}{2}} = 81$
   $4x + 3 = 6561$
   $4x = 6558$
   $x = 1639.5$
10. $(x - 15)^{\frac{3}{5}} = 343$
    $x = 16{,}807$
    $x = 16{,}822$
11. $x = 3$

Answer Key

Page 41

1. {-5, 5}
2. {-1, 5}
3. {-4, 4}
4. {-7, 7}
5. {-2, 2}
6. {-1, 6}
7. {-3, 1}
8. {-24, -10}
9. {-3}
10. {-6, -4}
11. {1, 5}
12. {-5, -2}

Page 42

1. {-2, 4}
2. {-10, -1}
3. $\{-1 \pm \sqrt{5}\}$
4. $\{\frac{-5 \pm \sqrt{53}}{2}\}$
5. $\{-1, 2\frac{1}{2}\}$
6. $\{\frac{-2 \pm \sqrt{6}}{2}\}$
7. $\{\frac{-2 \pm \sqrt{39}}{7}\}$
8. $\{\frac{-5 \pm \sqrt{10}}{3}\}$
9. $\{\frac{3 \pm \sqrt{41}}{4}\}$
10. $\{\frac{-7 \pm \sqrt{113}}{16}\}$
11. {0, 4}
12. $\{\frac{1 \pm \sqrt{33}}{4}\}$

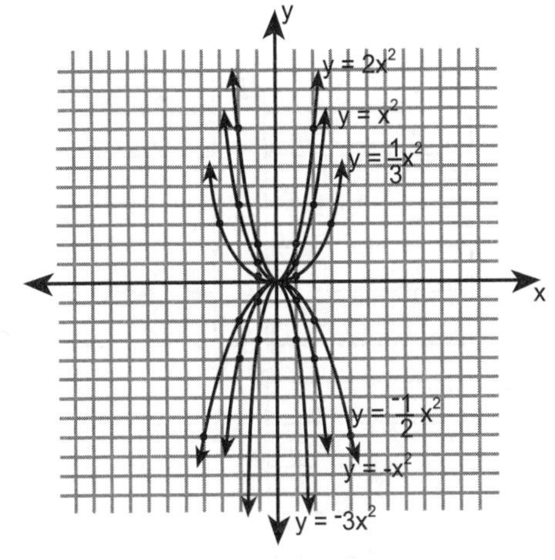

If $a > 1$, the graph narrows.
If $a < 1$, the graph widens.
$-a$ turns the graph upside down.

Page 43

1.

x	y
-2	8
-1	2
0	0
1	2
2	8

2.

x	y
-2	-4
-1	-1
0	0
1	-1
2	-4

3.

x	y
-2	-12
-1	-3
0	0
1	-3
2	-12

4.

x	y
-3	3
-1	1/3
0	0
1	1/3
3	3

5.

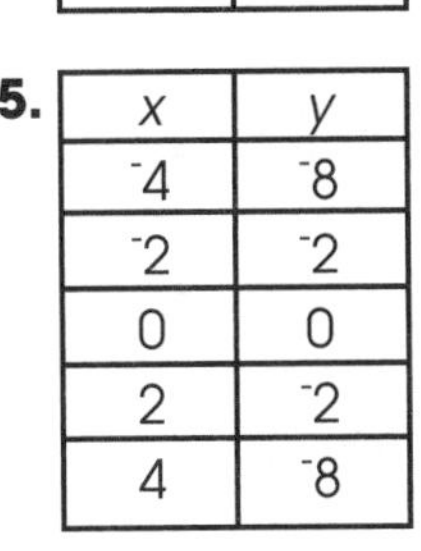

x	y
-4	-8
-2	-2
0	0
2	-2
4	-8

Page 44

1.

x	y
-2	2
-1	-1
0	-2
1	-1
2	2

2.

x	y
-2	8
-1	2
0	0
1	2
2	8

3.

x	y
-4	-1
-2	-7
0	-9
2	-7
4	-1

4.

x	y
-2	-6
-1	3
0	6
1	3
2	-6

5.

x	y
-3	3
-1	$-2\frac{1}{3}$
0	-3
1	$-2\frac{1}{3}$
3	3

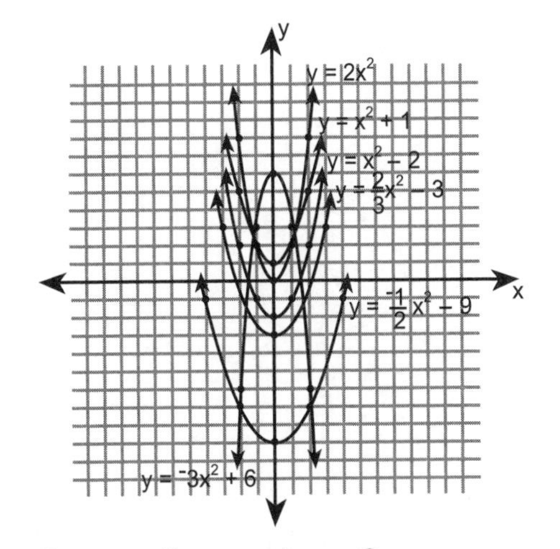

C moves the graph up C.
$-C$ moves the graph down C.

Answer Key

Page 45

1.

x	y
$^{-}5$	8
$^{-}4$	2
$^{-}3$	0
$^{-}2$	2
$^{-}1$	8

2.

x	y
2	3
3	$1\frac{1}{3}$
5	0
7	$1\frac{1}{3}$
8	3

3.

x	y
5	$^{-}8$
6	$^{-}2$
7	0
8	$^{-}2$
9	$^{-}8$

4.

x	y
$^{-}5$	1
$^{-}4$	$-\frac{1}{2}$
$^{-}3$	$^{-}1$
$^{-}2$	$-\frac{1}{2}$
$^{-}1$	1

5.

x	y
$^{-}10$	$^{-}3$
$^{-}9$	0
$^{-}8$	1
$^{-}7$	0
$^{-}6$	$^{-}3$

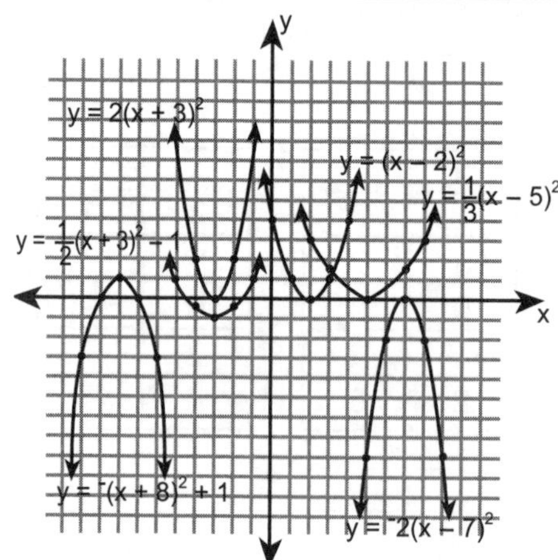

B moves the graph left *b* units.
$-B$ moves the graph right *b* units.

Page 46

1. $a = 1 \quad b = {}^{-}2 \quad c = 1$

$$\left(\frac{^{-}(^{-}2)}{2(1)}, \frac{^{-}(^{-}2)^2}{4(1)} + 1\right) = (1, 0)$$

2. $a = 8 \quad b = {}^{-}16 \quad c = 1$

$$\left(\frac{^{-}(^{-}16)}{2(8)}, \frac{^{-}(^{-}16)^2}{4(8)} + 1\right) = (1, {}^{-}7)$$

3. $a = {}^{-}3 \quad b = 6 \quad c = 1$

$$\left(\frac{^{-}(6)}{2(^{-}3)}, \frac{^{-}(6)^2}{4(^{-}3)} + {}^{-}1\right) = (1, 2)$$

4. $a = 1 \quad b = 0 \quad c = 9$

$$\left(\frac{^{-}0}{2(1)}, \frac{^{-}(0)^2}{4(1)} + {}^{-}9\right) = (0, {}^{-}9)$$

5. $a = 2 \quad b = 0 \quad c = 1$

$$\left(\frac{^{-}0}{4(2)}, \frac{^{-}(0)^2}{4(2)} + 1\right) = (0, 1)$$

6. $a = 1 \quad b = 0 \quad c = 0 \quad (0, 0)$

7. $a = {}^{-}10 \quad b = 0 \quad c = 0 \quad (0, 0)$

8. $a = 2 \quad b = 6 \quad c = 0$

$$\left(\frac{^{-}6}{2(2)}, \frac{^{-}(6)^2}{4(2)} + 0\right) = \left(\frac{^{-}3}{2}, \frac{^{-}9}{2}\right)$$

9. $a = {}^{-}4 \quad b = 0 \quad c = {}^{-}7$

$$\left(\frac{^{-}0}{2(^{-}4)}, \frac{^{-}(0)^2}{4(^{-}4)} + {}^{-}7\right) = (0, {}^{-}7)$$

10. It is the C term.

Answer Key

Page 47

1. $\dfrac{{}^{-}({}^{-}14) \pm \sqrt{({}^{-}14)^2 - 4(3)(1)}}{2(3)} = \dfrac{14 \pm \sqrt{184}}{6}$
$= \dfrac{7 \pm \sqrt{46}}{3}$

2. $\dfrac{{}^{-}({}^{-}1) \pm \sqrt{({}^{-}1)^2 - 4(2)({}^{-}1)}}{2(2)} = \dfrac{1 \pm \sqrt{9}}{4}$
$= \dfrac{1 \pm 3}{4} = 1 \text{ and } \dfrac{{}^{-}1}{2}$

3. $\dfrac{{}^{-}({}^{-}6) \pm \sqrt{36 - 4({}^{-}3)(1)}}{2({}^{-}3)} = \dfrac{6 \pm \sqrt{48}}{{}^{-}6}$
$= \dfrac{6 \pm 4\sqrt{3}}{{}^{-}6} = \dfrac{3 \pm 2\sqrt{3}}{{}^{-}3}$

4. $\dfrac{{}^{-}1 \pm \sqrt{1 - 4(3)(0)}}{2(3)} = \dfrac{{}^{-}1 \pm 1}{6}$
$= 0 \text{ and } \dfrac{{}^{-}1}{3}$

5. $\dfrac{14 \pm \sqrt{({}^{-}14)2 - 4({}^{-}7)^0}}{{}^{-}14} = 14 \pm \sqrt{196 - 0}$
$= \dfrac{14 \pm 14}{{}^{-}14} = {}^{-}2, 0$

6. $\dfrac{{}^{-}0 \pm \sqrt{0^2 - 4(4)({}^{-}9)}}{2(4)} = \dfrac{\pm\sqrt{144}}{8}$
$= \dfrac{\pm 12}{8} = \dfrac{3}{2} \text{ and } \dfrac{{}^{-}3}{2}$

7. $\dfrac{{}^{-}0 \pm \sqrt{0^2 - 4(1)({}^{-}2)}}{2(1)} = \dfrac{\pm\sqrt{8}}{2}$
$= \dfrac{\pm 2\sqrt{2}}{2} = \sqrt{2} \text{ and } {}^{-}\sqrt{2}$

8. $x^2 - x = 6$
$= \dfrac{1 \pm \sqrt{1 - (4)(1)({}^{-}6)}}{2} = \dfrac{1 \pm \sqrt{25}}{2}$
$= \dfrac{1 \pm 5}{2} = 3 \text{ and } {}^{-}2$

9. $2x^2 + 6x + 0 = \dfrac{{}^{-}6 \pm \sqrt{36}}{4}$
$= \dfrac{{}^{-}6 \pm 6}{4} = 0 \text{ and } {}^{-}3$

Page 48

1. $y = (x - 6)(x - 2)$
$x = 6, 2$
$v = \left(\dfrac{8}{2}, \dfrac{{}^{-}64}{4} + 12\right)$
$= (4, {}^{-}4)$

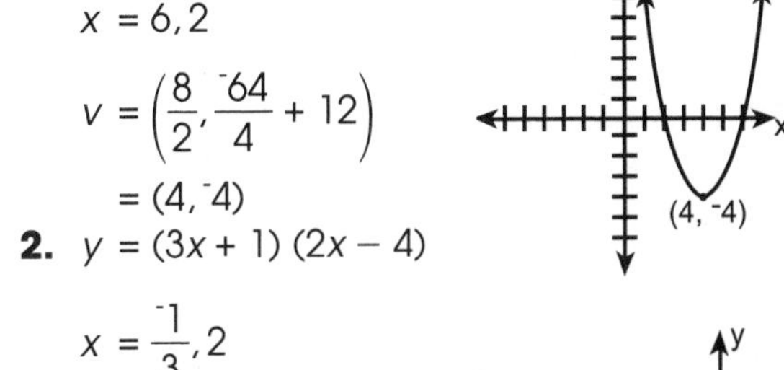

2. $y = (3x + 1)(2x - 4)$
$x = \dfrac{{}^{-}1}{3}, 2$
$v = \left(\dfrac{10}{12}, \dfrac{{}^{-}100}{24} - 4\right)$
$= \left(\dfrac{5}{6}, \dfrac{{}^{-}196}{24}\right) = \left(\dfrac{5}{6}, \dfrac{{}^{-}49}{6}\right)$

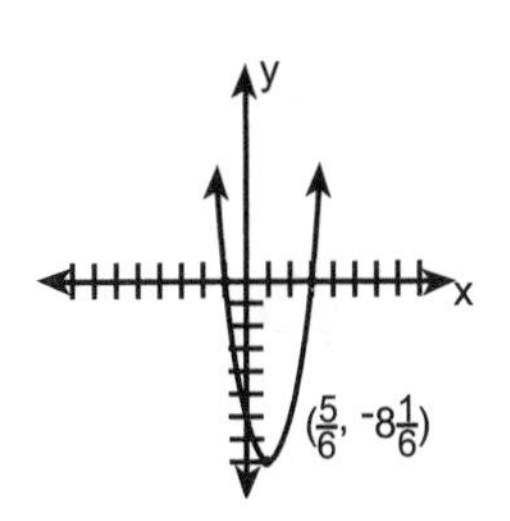

3. $y = ({}^{-}5x + 2)(3x + 1)$
$x = \dfrac{2}{5}, \dfrac{{}^{-}1}{3}$
$v = \left(\dfrac{1}{30}, \dfrac{121}{60}\right)$

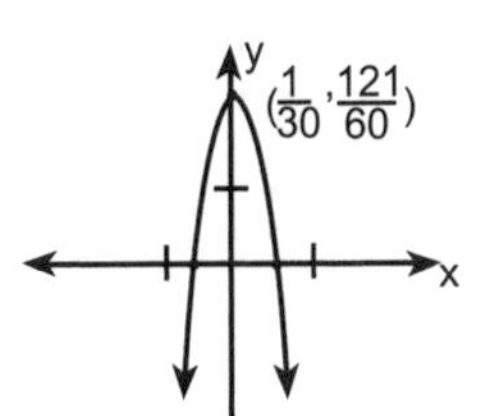

4. $x = {}^{-}4, 4$
$v = (0, {}^{-}16)$

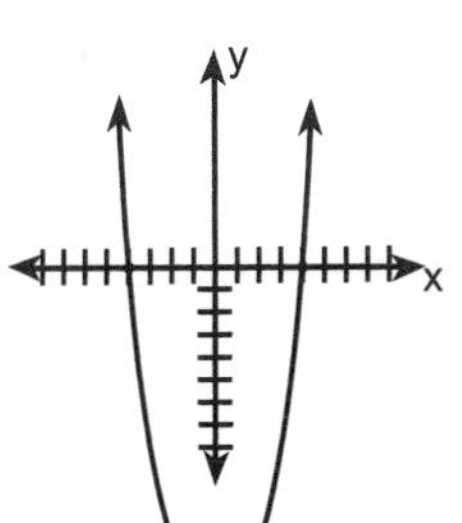

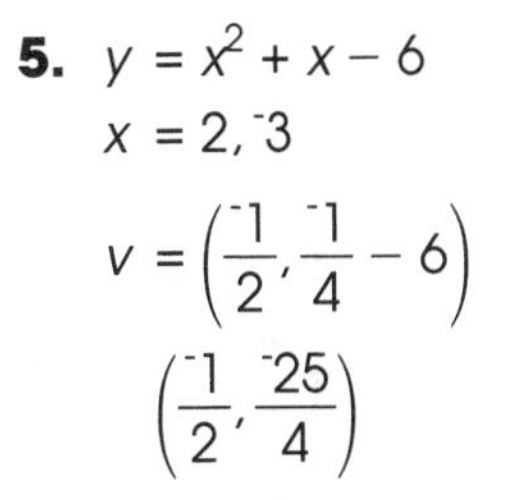

5. $y = x^2 + x - 6$
$x = 2, {}^{-}3$
$v = \left(\dfrac{{}^{-}1}{2}, \dfrac{{}^{-}1}{4} - 6\right)$
$\left(\dfrac{{}^{-}1}{2}, \dfrac{{}^{-}25}{4}\right)$

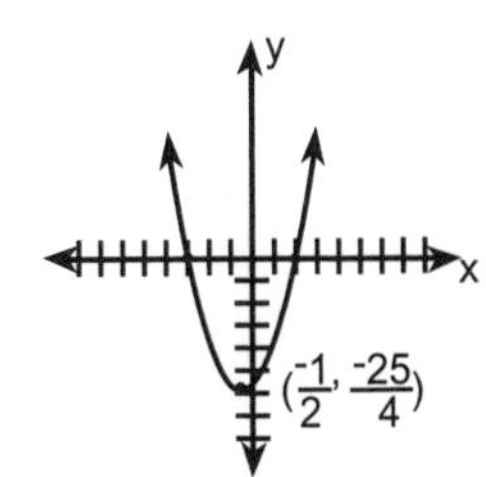

6. $y = x^2 - 25$
$x = \pm 5$
$v = (0, {}^{-}25)$

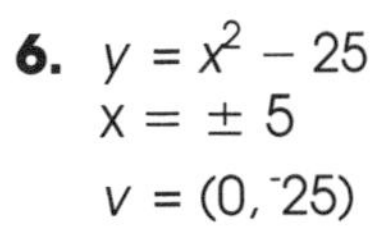

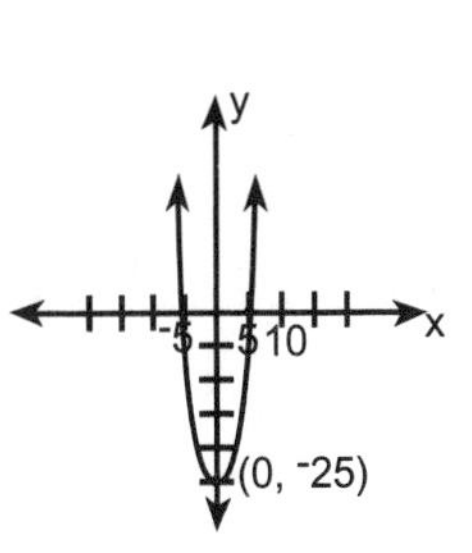

7. $y = (x + 3)(x + 2)$

$x = {}^{-}3 \qquad x = {}^{-}2$

$v = \left(\frac{{}^{-}5}{2}, \frac{25}{4} - 6\right)$

$\left(\frac{{}^{-}5}{2}, \frac{1}{4}\right)$

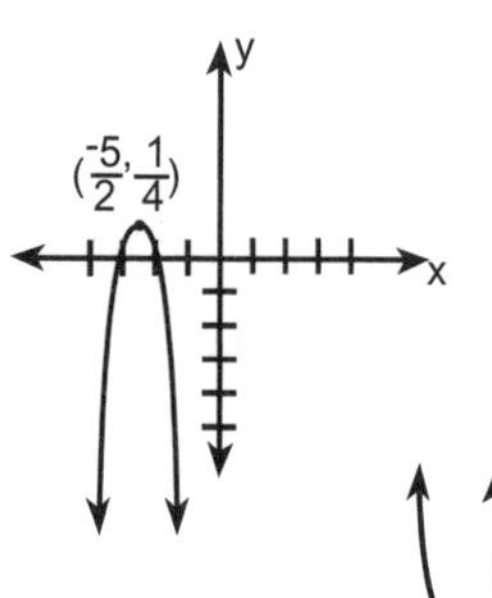

8. $x = \pm 0$

$v = (0, 0)$

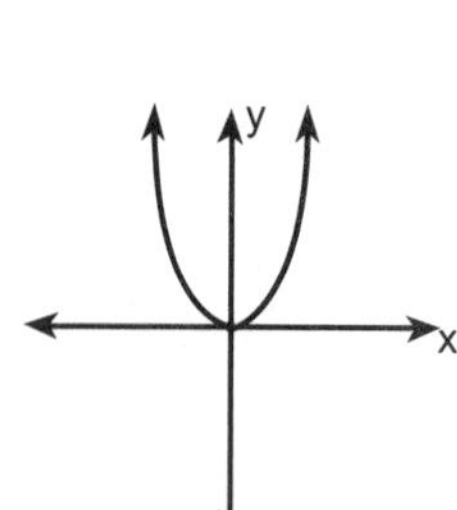

Page 49

1.

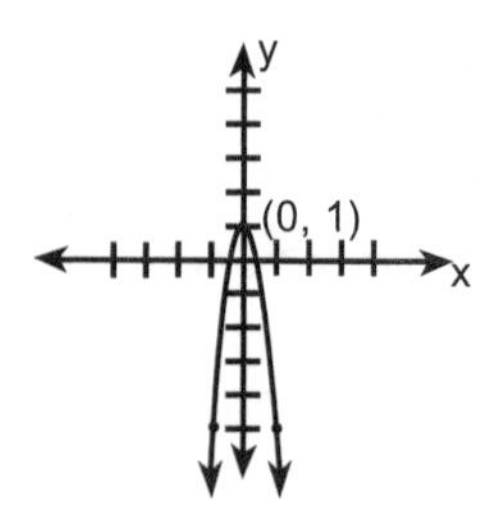

2.

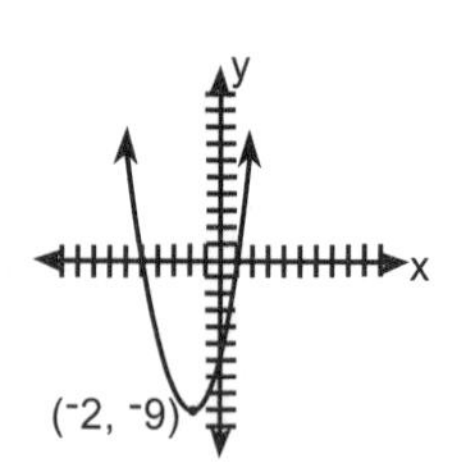

3.

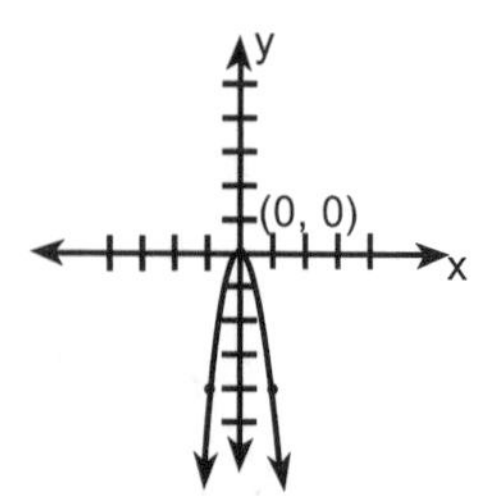

4.

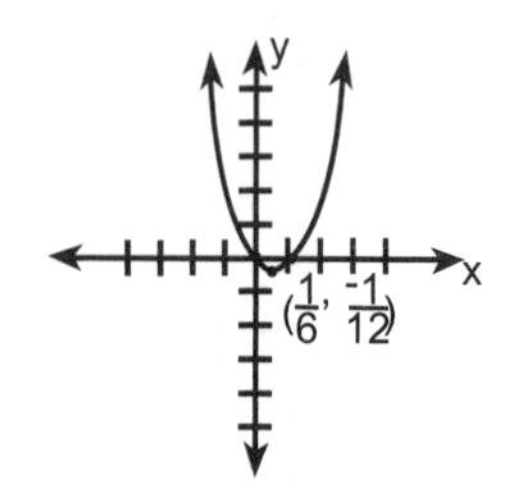

5.

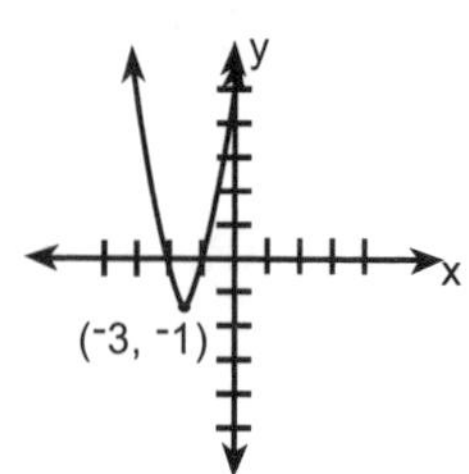

6.

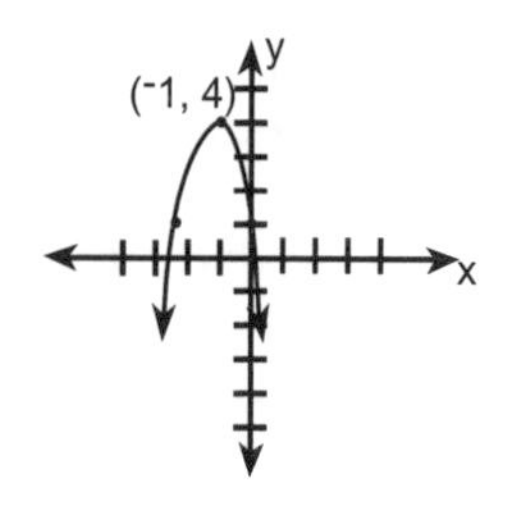

7.

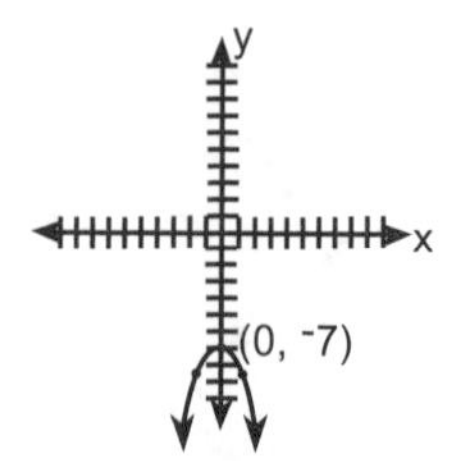

8.

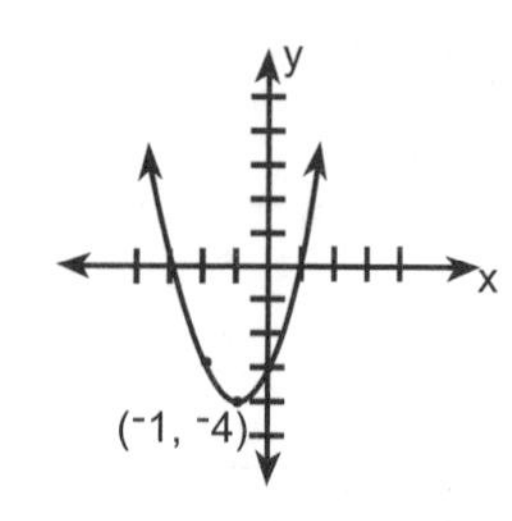

9.

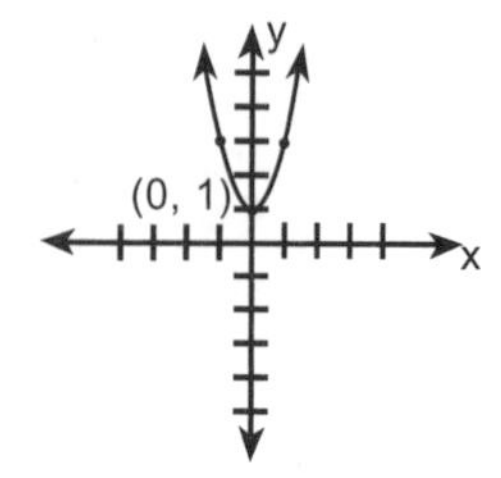

10.

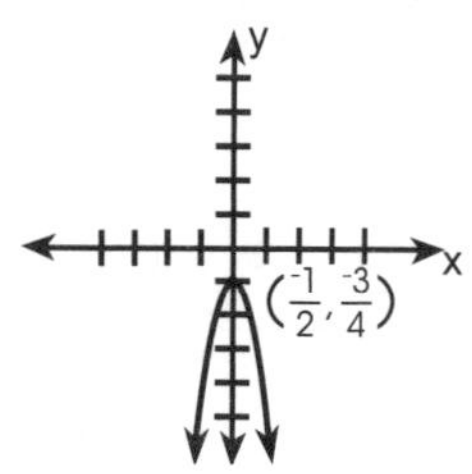

Page 50

	brown	white	blue	green	Bear	Bull	Whale	Gorilla	first	second	third	fourth
Billy Brown				X			X			X		
Willy White			X			X					X	
Bobby Blue	X							X	X			
George Green		X			X							X

Page 51

1. $\log 1000 = y$
$10^y = 1000$
$y = 3$

2. $\log \sqrt[5]{10} = y$
$10^y = 10^{\frac{1}{5}}$
$y = \frac{1}{5}$

3. $\log \sqrt[3]{10^2} = y$
$10^y = 10^{\frac{2}{3}}$
$y = \frac{2}{3}$

4. $\log 0.1 = y$
$10^y = 0.1$
$y = {}^{-}1$

5. $\log 0.0001 = y$
$10^y = 0.0001$
$y = {}^{-}4$

6. $\log \sqrt[4]{10} = y$
$10^y = 10^{\frac{1}{4}}$
$y = \frac{1}{4}$

7. $\log \sqrt{10} = y$
$10^y = 10^{\frac{1}{2}}$
$y = \frac{1}{2}$

8. $\log 10^6 = y$
$10^y = 10^6$
$y = 6$

9. $\log 1 = y$
$10^y = 1$
$y = 0$

10. $\log 10{,}000 = y$
$10^y = 10{,}000$
$y = 4$

Answer Key

Page 52

1. $\log_5 125 = 3$
2. $\log_{10} 1{,}000{,}000 = 6$
3. $10^0 = 1$
4. $3^{-5} = \frac{1}{243}$
5. $\log_7 16{,}807 = 5$
6. $10^y = x$
7. $\log_{12} 87 = x$
8. $15^y = 30$
9. $Q^y = x$
10. $180^y = B$
11. $\log_{10} x = y$
12. $b^3 = 64$
13. $x^{10} = 5$
14. $\log_7 343 = x$

Page 53

1.

x	f(x)
1/9	-2
1/3	-1
1	0
3	1
9	2
27	3

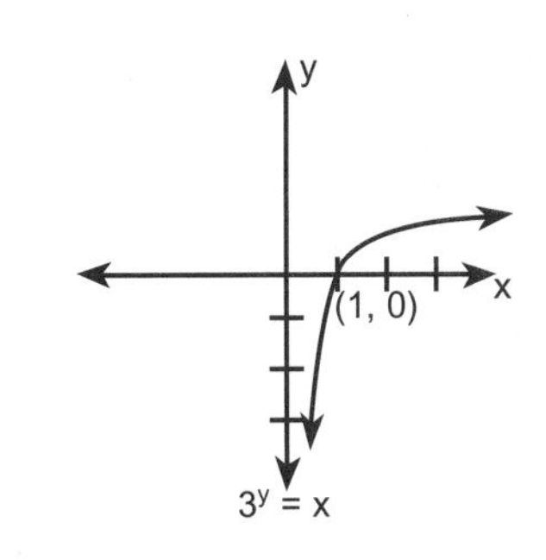

2.

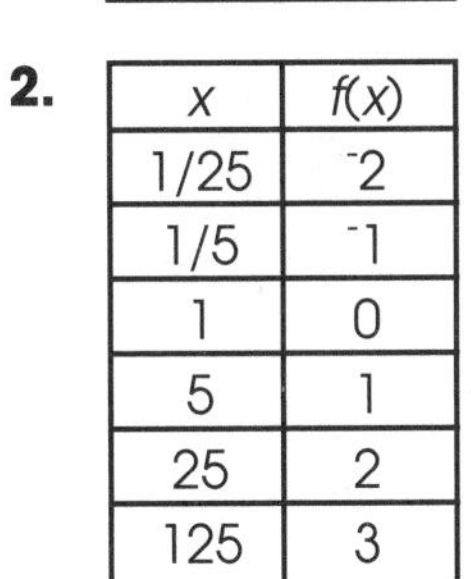

x	f(x)
1/25	-2
1/5	-1
1	0
5	1
25	2
125	3

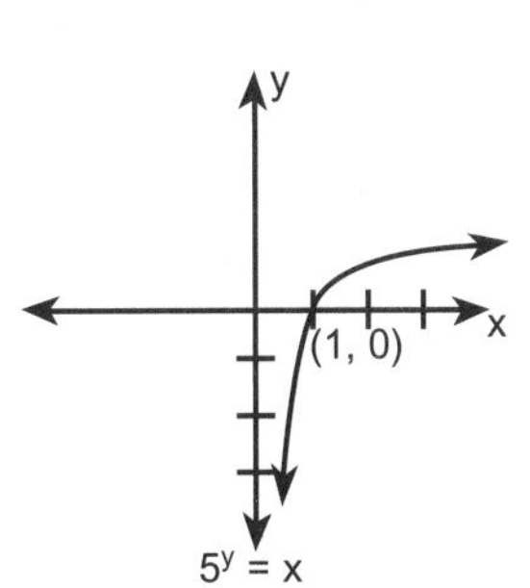

3.

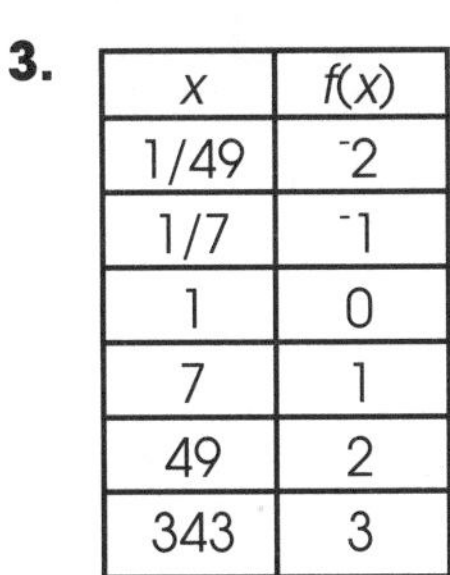

x	f(x)
1/49	-2
1/7	-1
1	0
7	1
49	2
343	3

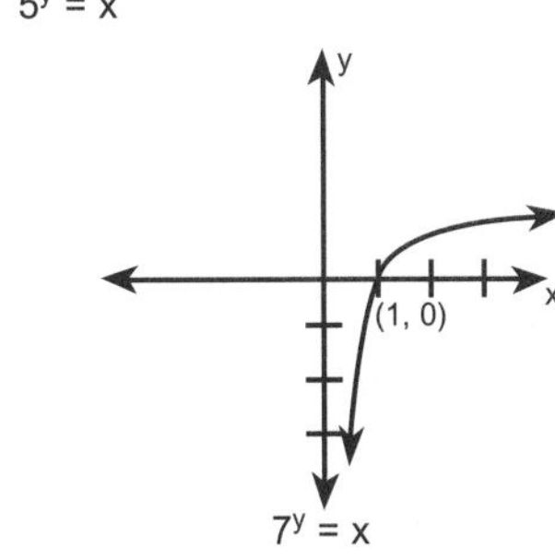

4.

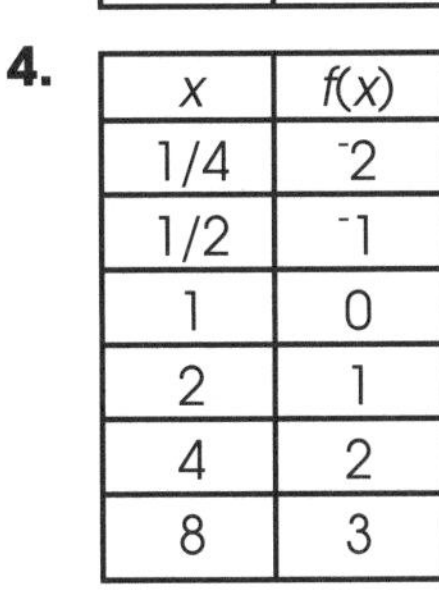

x	f(x)
1/4	-2
1/2	-1
1	0
2	1
4	2
8	3

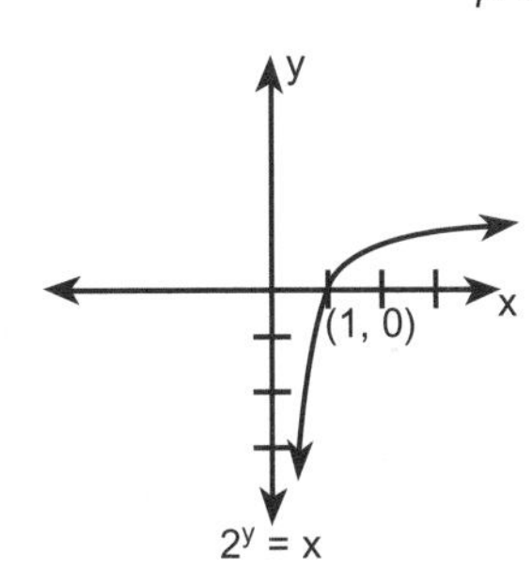

5.

x	f(x)
1/16	-2
1/4	-1
1	0
4	1
16	2
64	3

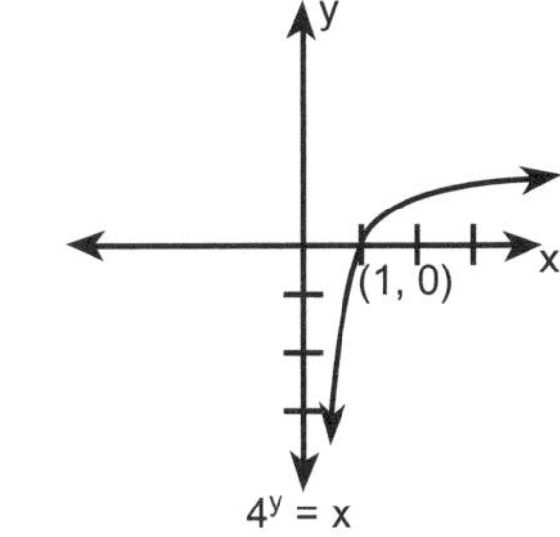

6.

x	f(x)
1/121	-2
1/11	-1
1	0
11	1
121	2
1331	3

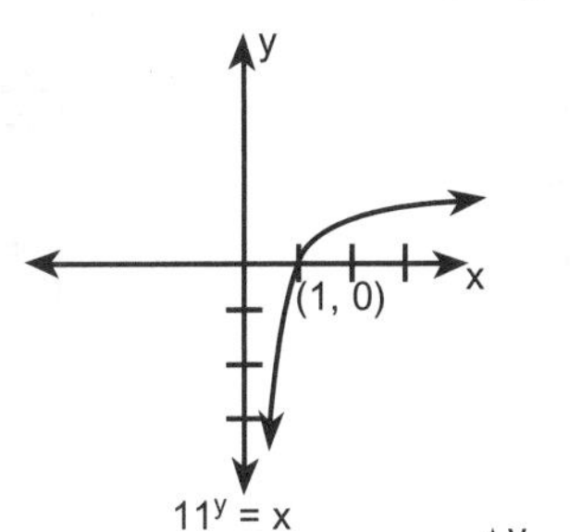

7.

x	f(x)
1/225	-2
1/15	-1
1	0
15	1
225	2
3375	3

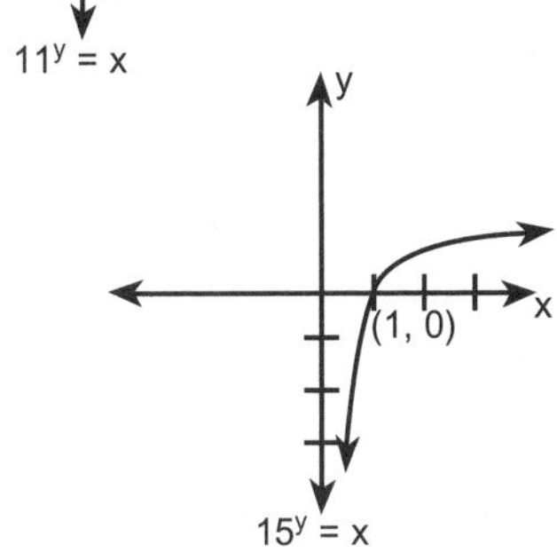

Page 54

1. $\log_9 24$
2. $\log_{12} 132$
3. $\log_{16} 3$
4. $\log\left(\frac{3}{2}\right)$
5. $6 \log 14$
6. $16 \log_{20} 10$
7. Cannot simplify because of different bases.
8. $\log 100$
9. $\log 5^3 = 3 \log 5$
10. $4 \log_2 2$

Answer Key

Page 55

1. $\log_3 x - \log_3 2^2 = \log_3 3^3$
   $\log_3 \left(\frac{x}{4}\right) = \log_3 27$
   $\frac{x}{4} = 27 \quad x = 108$

2. $2^9 = x$
   $x = 512$

3. $2^x = 128$
   $x = 7$

4. $x^{-2} = 144$
   $\frac{1}{x^2} = 144$
   $x = \frac{1}{12}$

5. $\log_2 x = \log_2 27^{\frac{1}{3}}$
   $\log_2 x = \log 3$
   $x = 3$

6. $\log_{16} 16 = x$
   $16^x = 16$
   $x = 1$

7. $\log 2^5 = \log x$
   $x = 2^5 = 32$

8. $\log_2 \frac{x}{5} = \log_2 10$
   $\frac{x}{5} = 10$
   $x = 50$

A logarithm is an $\frac{E}{2}\ \frac{X}{6}\ \frac{P}{4}\ \frac{O}{7}\ \frac{N}{5}\ \frac{E}{3}\ \frac{N}{1}\ \frac{T}{8}$.

Page 56

1. 1.5
2. 3
3. No solution
4. 2.7
5. 2.3
6. 3.1
7. 6.5
8. 4.5
9. 1.5
10. 2.1

Page 57

1. Domain: $\mathbb{R}$; Range: $\mathbb{R}$
2. Domain: $\mathbb{R}$; Range: $y \leq 1$
3. Domain: $\mathbb{R}$; Range: $y = 1$
4. Domain: $\mathbb{R}$; Range: $\mathbb{R}$
5. Domain: $\mathbb{R}$; Range: $y \leq 0$
6. Domain: $\mathbb{R}$; Range: $\mathbb{R}$
7. Domain: $\mathbb{R}$; Range: $y \geq {}^{-}2$
8. Domain: $\mathbb{R}$; except 0; Range: $y > 0$
9. Domain: $\mathbb{R}$; except 0; Range: $\mathbb{R}$, except 0

Page 58

1. Domain: ${}^{-}7 \leq x \leq 7$; Range: $y = 3$
2. Domain: ${}^{-}6 \leq x \leq 6$; Range: ${}^{-}3 \leq y \leq 3$
3. Domain: ${}^{-}7 \leq x \leq 7$; Range: ${}^{-}5 \leq y \leq 4$
4. Domain: ${}^{-}6 \leq x \leq 6$; Range: ${}^{-}3 \leq y \leq 2$
5. Domain: ${}^{-}4 \leq x \leq 4$; Range: ${}^{-}7 \leq y \leq 7$
6. Domain: ${}^{-}4 \leq x \leq 4$; Range: ${}^{-}7 \leq y \leq 7$
7. Domain: ${}^{-}6 \leq x \leq 6$; Range: ${}^{-}3 \leq y \leq 2$
8. Domain: ${}^{-}3 \leq x \leq 3$; Range: ${}^{-}4 \leq y \leq 7$
9. Domain: ${}^{-}7 \leq x \leq 7$; Range: ${}^{-}3 \leq y \leq 3$

Page 59

	Salesperson	Pharmacist	Grocer	Police Officer
Clark		X		
Jones			X	
Morgan	X			
Smith				X

Page 60

1. $\sin B = \frac{3}{5}$ $\quad \cos B = \frac{4}{5}$ $\quad \tan B = \frac{3}{4}$
2. $\sin A = \frac{15}{39}$ $\quad \cos A = \frac{36}{39}$ $\quad \tan A = \frac{15}{36}$
   $\sin B = \frac{36}{39}$ $\quad \cos B = \frac{15}{39}$ $\quad \tan B = \frac{36}{15}$
3. $\sin A = \frac{Y}{Z}$ $\quad \cos A = \frac{X}{Z}$ $\quad \tan A = \frac{Y}{X}$
   $\sin B = \frac{X}{Z}$ $\quad \cos B = \frac{Y}{Z}$ $\quad \tan B = \frac{X}{Y}$
4. $\sin A = \frac{12}{13}$ $\quad \cos A = \frac{5}{13}$ $\quad \tan A = \frac{12}{5}$
   $\sin B = \frac{5}{13}$ $\quad \cos B = \frac{12}{13}$ $\quad \tan B = \frac{5}{12}$

They always have the same ratio.

Page 61

1. = 1
2. = 0.98
3. = ${}^{-}0.77$
4. = 0.98
5. = 0.39
6. = ${}^{-}1$
1. = 68°
2. = 48°
3. = 72°
4. = 54°
5. = 75
6. = 5°

Page 62

1. $x = 11.3, \quad y = 4.1$
2. $x = 38.6, \quad y = 37.3$
3. $x = 9.1, \quad y = 4.3$
4. $x = 14, \quad y = 12.1$
5. $x = 21.9, \quad y = 11.0$
6. $x = 19.1, \quad y = 19.1$
7. $x = 8.6, \quad y = 12.3$
8. $x = 4.9, \quad y = 28.4$

Page 63

1. $x = 30°, \quad y = 60°$
2. $x = 45°, \quad y = 45°$
3. $x = 62°, \quad y = 28°$
4. $x = 60°, \quad y = 30°$
5. $x = 62°, \quad y = 28°$
6. $x = 13°, \quad y = 77°$
7. $x = 71°, \quad y = 19°$
8. $x = 57°, \quad y = 33°$

Answer Key

Page 64

1. $\cos x = \frac{3}{20}$, $x = 81.4°$

2. $\tan 30° = \frac{n}{100}$, $h = 57.7 + 6 = 63.7$ ft.

3. $\tan x = \frac{1500}{9000} = 9.5°$

4. $\sin x = \frac{30}{150} = 12°$

5. $\sin 30° = \frac{15}{x}$ $x = 30$ ft.

Page 66

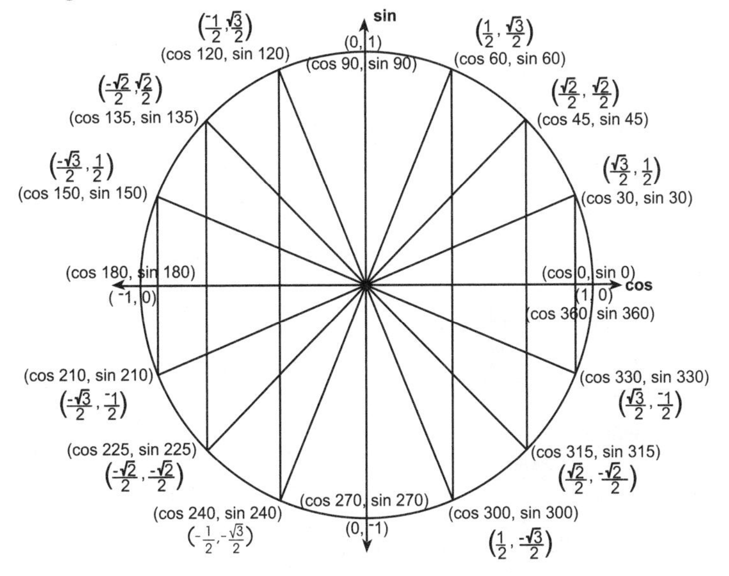

Page 67

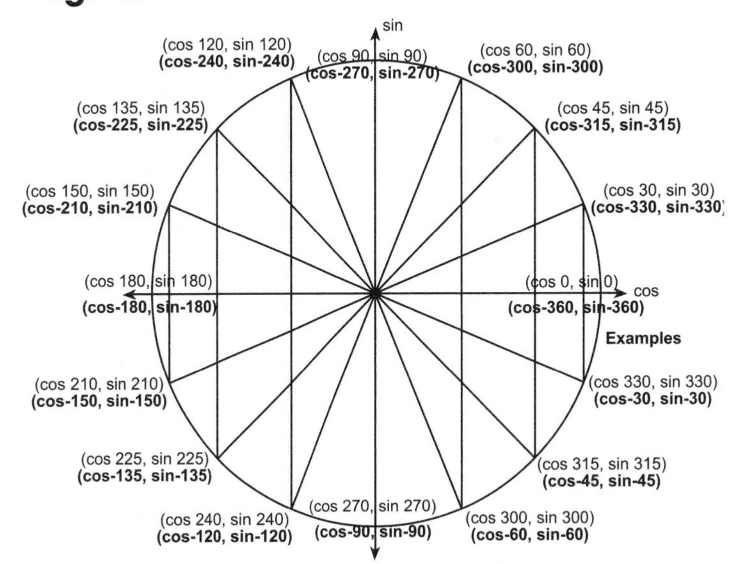

The absolute value of the negative measure plus the positive angle measure equals 360°.

Page 68

1. $630 - 360 = 270 - 360 = \sin - 90 = {}^{-}1$
2. ${}^{-}750 - 360 = 390 - 360 = \sin - 30 = -\frac{1}{2}$
3. ${}^{-}480 - 360 = 120 - 360 = \cos - 240 = -\frac{1}{2}$
4. ${}^{-}420 - 360 = 60 - 360 = \sin - 300 = -\frac{\sqrt{3}}{2}$
5. $510 - 360 = 150 - 360 = \sin - 210 = \frac{1}{2}$
6. $1020 - 360 = 660 - 360 = \cos 300 = \frac{1}{2}$
7. ${}^{-}540 - 360 = 900 - 360 = 540 - 360 =$ $180 - 360 = {}^{-}180 = {}^{-}1$
8. ${}^{-}675 + 360 = 315 - 360 = -45 = \frac{\sqrt{2}}{2}$
9. ${}^{-}540 - 360 = 90 - 360 = {}^{-}270 - 360 = {}^{-}90 = 0$
10. ${}^{-}930 - 360 = 570 - 360 = 210 - 360 = {}^{-}150 = -\frac{\sqrt{3}}{2}$
11. $405 - 360 = 45 = \frac{\sqrt{2}}{2}$
12. ${}^{-}600 - 360 = 240 - 360 = {}^{-}120 + 300 = \frac{\sqrt{3}}{2}$
13. $3600 - 360 = 3240 - 360 = 2880 \times 10 = 0$
14. ${}^{-}1830 + 360 = 1470 - 360 = 1110 - 360 =$ $750 - 360 = 390 - 360 = 30 = \frac{\sqrt{3}}{2}$

Answer Key

List the pairs: 1 & 7; 2 & 3; 4 & 10; 5 & 6; 9 & 13; 11 & 8; 12 & 14

In each pair, what is the relationship of the reference angles? The reference angles are complementary.

Page 69

1. $\frac{31}{18}\pi$ **2.** $\frac{5}{6}\pi$ **3.** $\frac{1}{6}\pi$
4. $\frac{7}{3}\pi$ **5.** $\frac{2}{3}\pi$ **6.** $\frac{35}{18}\pi$

1. 225° **2.** 720° **3.** 210°
4. 30° **5.** 315° **6.** 810°

Page 70

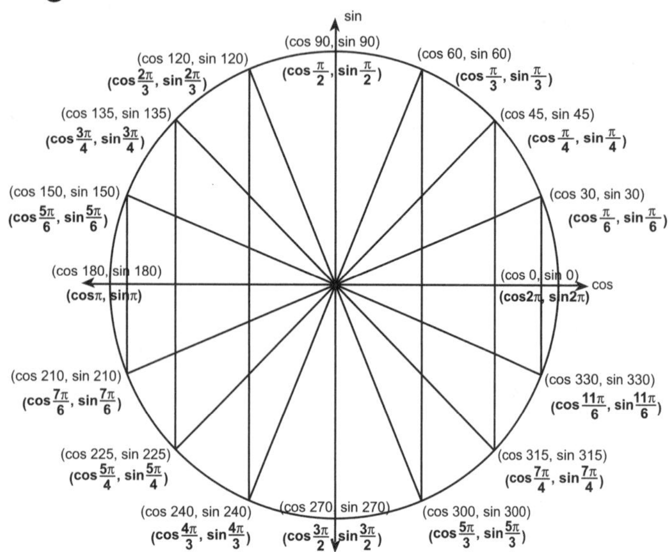

Page 71

Answers may vary.
1. cos 30°, cos 330° **2.** cos 75°, cos 285°
3. cos -90°, cos 450° **4.** cos 135°, cos 225°
5. cos 210°, cos 150°

Answers may vary.
1. sin 60°, sin 120° **2.** sin 150°, sin 30°
3. sin 0°, sin 180° **4.** sin 315°, sin -225°
5. sin 40°, sin 140°

Page 72

1.

x°	y
0	1
30	0.86
45	7
60	1/2
90	0
135	-0.7
180	-1
270	0
360	1

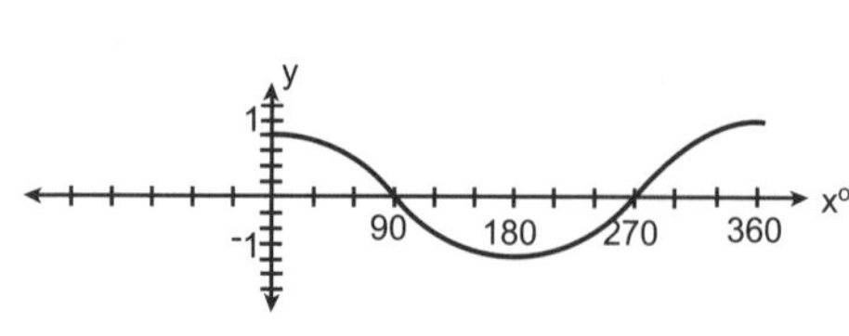

2.

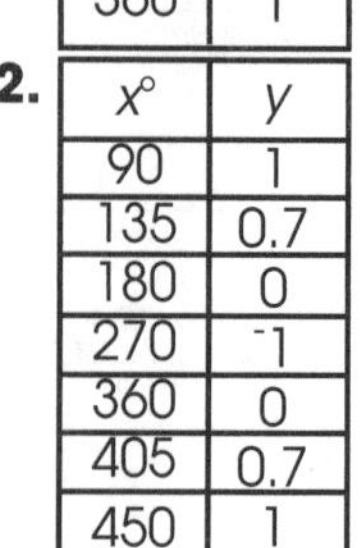

x°	y
90	1
135	0.7
180	0
270	-1
360	0
405	0.7
450	1

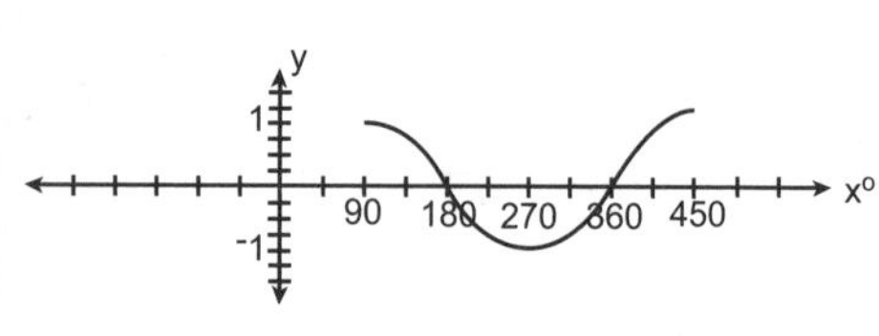

3.

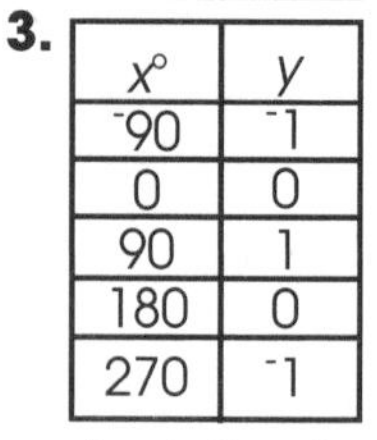

x°	y
-90	-1
0	0
90	1
180	0
270	-1

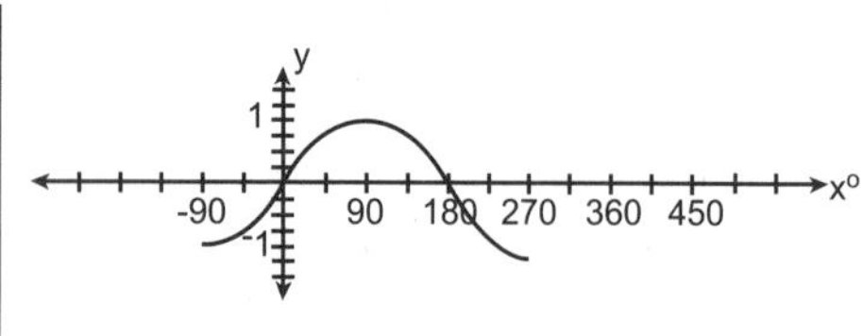

They look like the same graph shifted 90° to the right.

Page 73

4.

x°	y
-360	0
-270	1
-180	0
-90	-1
0	0
90	1
180	0
270	-1
360	0

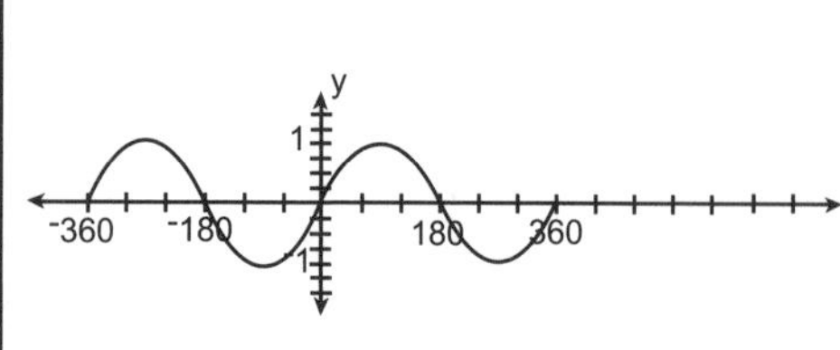

5.

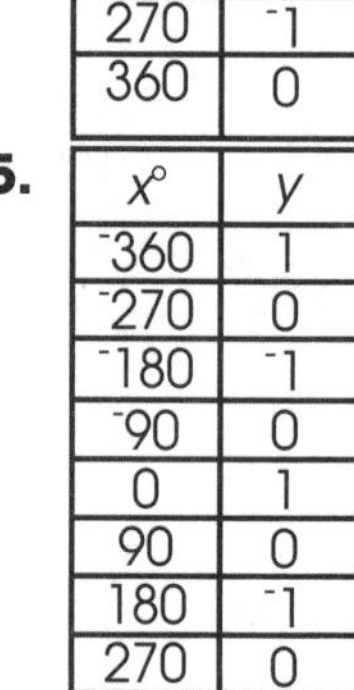

x°	y
-360	1
-270	0
-180	-1
-90	0
0	1
90	0
180	-1
270	0
360	1

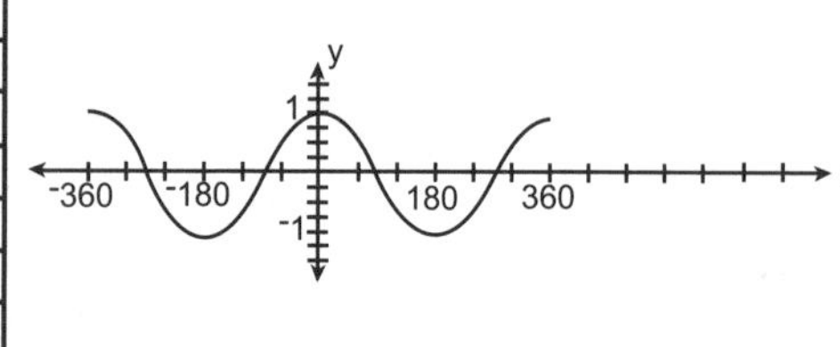

Answer Key

6.

x°	y
0	0
90	1
180	0
270	-1
360	0
450	1
540	0
630	-1
720	0

7.

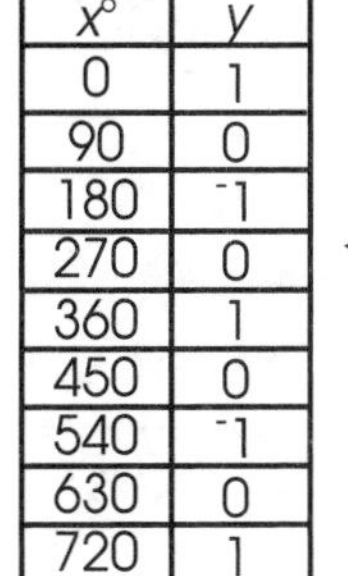

x°	y
0	1
90	0
180	-1
270	0
360	1
450	0
540	-1
630	0
720	1

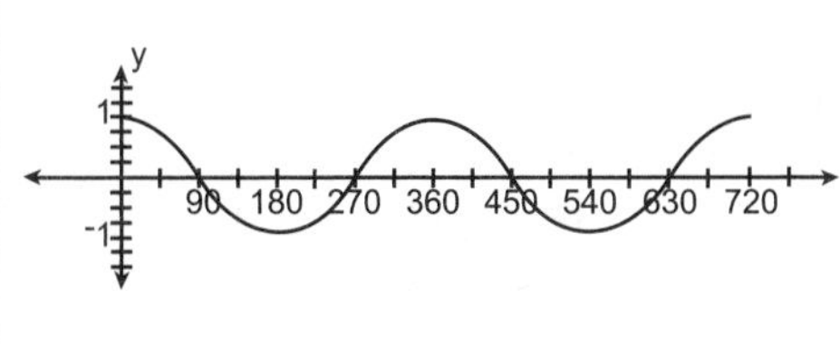

Page 74

1.

x°	y
0	4
90	0
180	-4
270	0
360	4

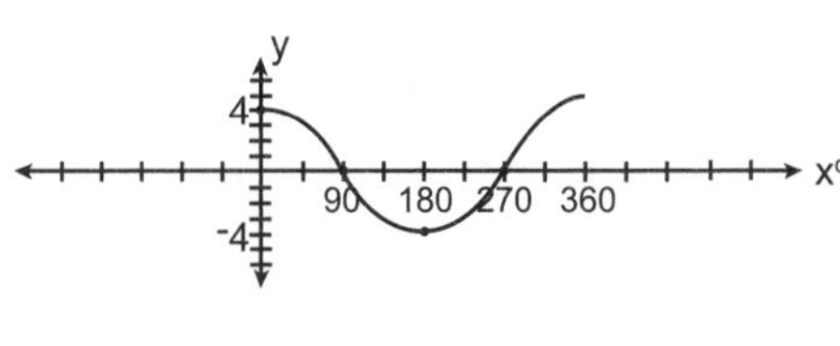

2.

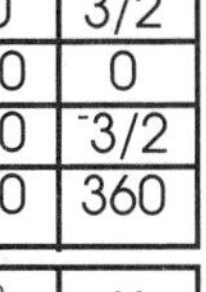

x°	y
0	0
90	3/2
180	0
270	-3/2
360	360

3.

x°	y
0	1/2
90	0
180	-1/2
270	0
360	1/2

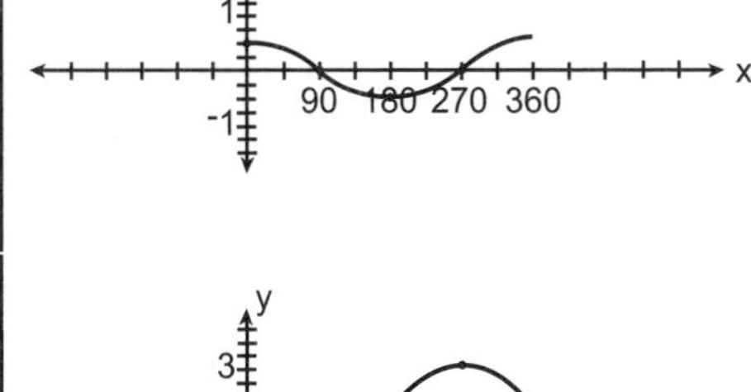

4.

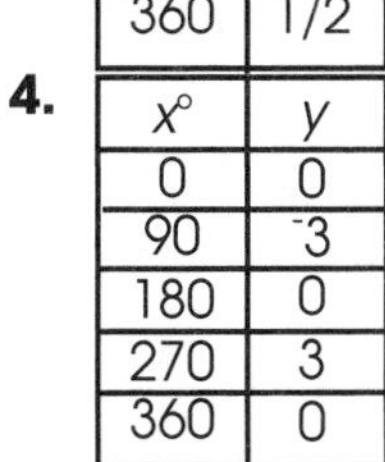

x°	y
0	0
90	-3
180	0
270	3
360	0

5.

x°	y
0	0
90	4
180	0
270	-4
360	0

6.

x°	y
0	0
90	-1/3
180	0
270	1/3
360	0

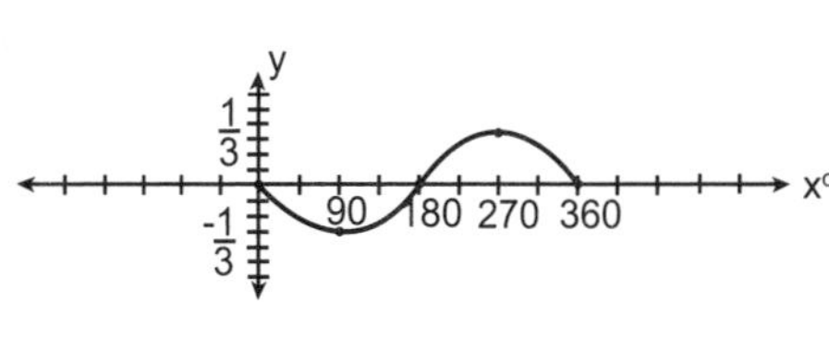

A increases the amplitude. -*A* inverts the graph.

Page 75

1.

x°	y
0	4
90	3
180	2
270	3
360	4

2.

x°	y
0	2
90	3
180	2
270	1
360	2

3.

x°	y
0	3/2
90	1/2
180	-1/2
270	1/2
360	3/2

4.

x°	y
0	-3
90	-2
180	-3
270	-4
360	-3

5.

x°	y
0	0
90	-1
180	-2
270	-1
360	0

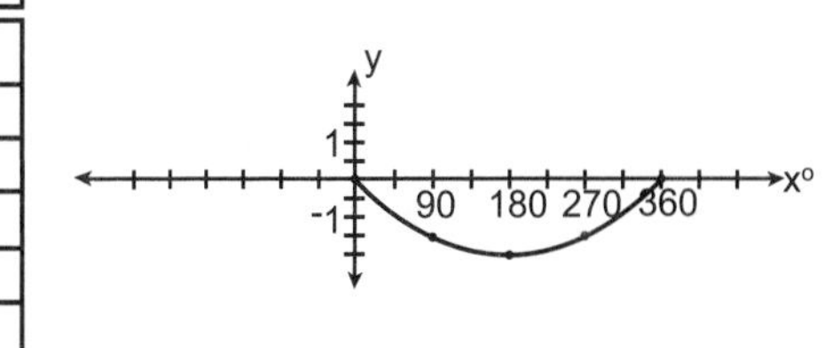

Answer Key

6.

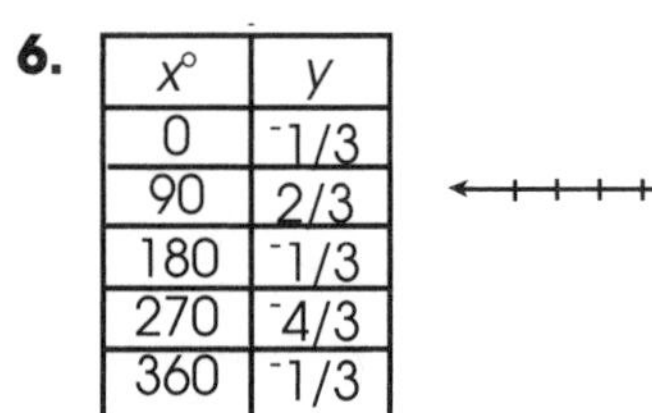

$x°$	y
0	-1/3
90	2/3
180	-1/3
270	-4/3
360	-1/3

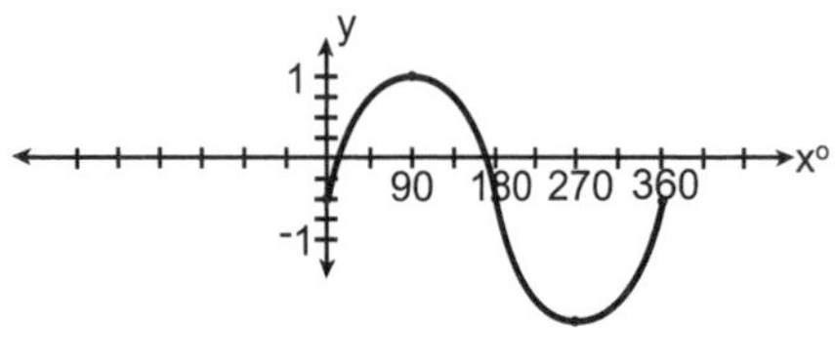

C moves the entire graph up. $-C$ moves the entire graph down.

Page 76

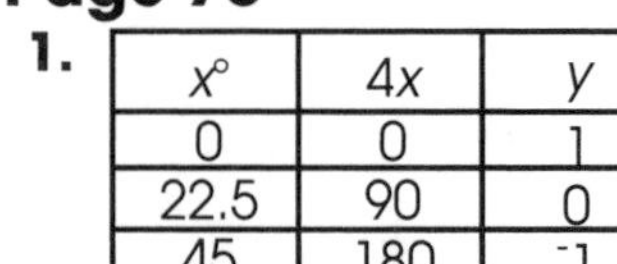

1.

$x°$	$4x$	y
0	0	1
22.5	90	0
45	180	-1
67.5	270	0
90	360	1

2.

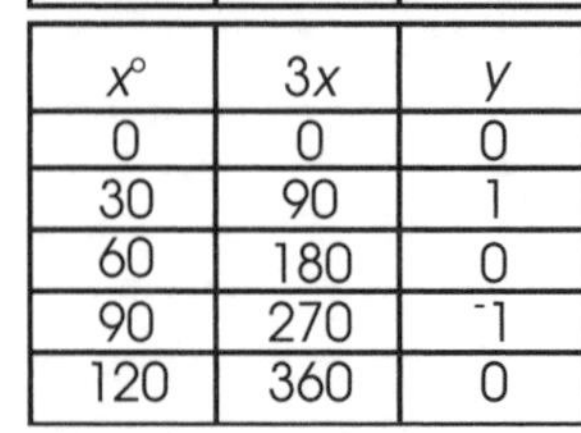

$x°$	$3x$	y
0	0	0
30	90	1
60	180	0
90	270	-1
120	360	0

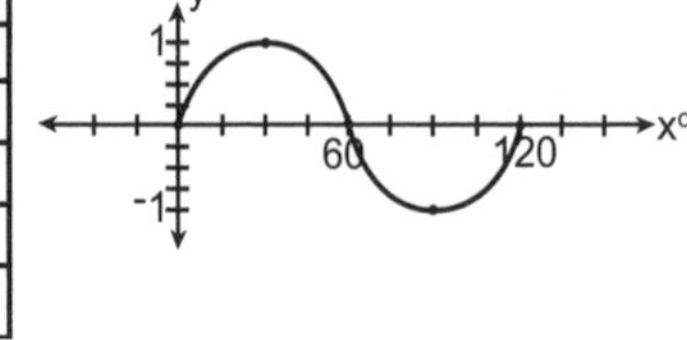

3.

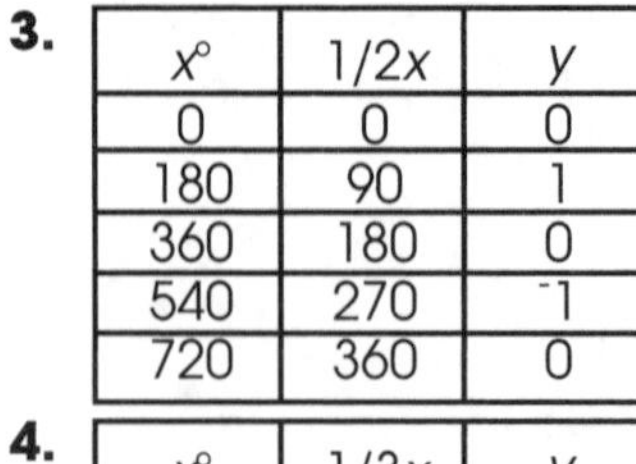

$x°$	$1/2x$	y
0	0	0
180	90	1
360	180	0
540	270	-1
720	360	0

4.

$x°$	$1/3x$	y
0	0	1
270	90	0
540	180	-1
810	270	0
1080	360	1

5.

$x°$	$1/4x$	y
0	0	0
360	90	1
720	180	0
1080	270	-1
1440	360	0

6.

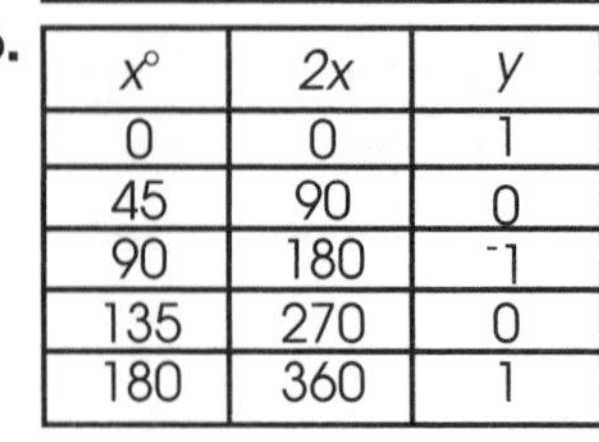

$x°$	$2x$	y
0	0	1
45	90	0
90	180	-1
135	270	0
180	360	1

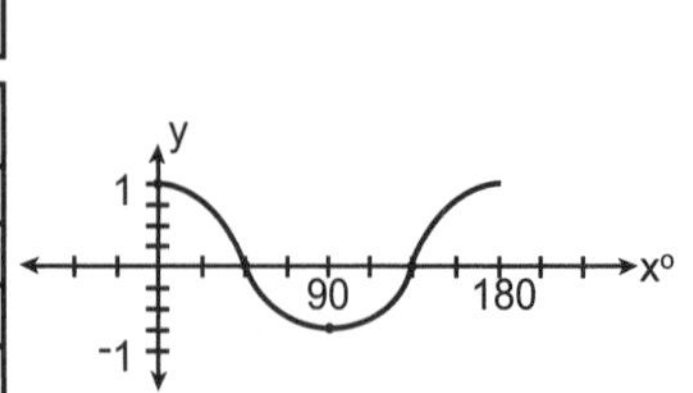

Page 77

1.

$x°$	$x - 90$	y
90	0	1
180	90	0
270	180	-1
360	270	0
450	360	1

2.

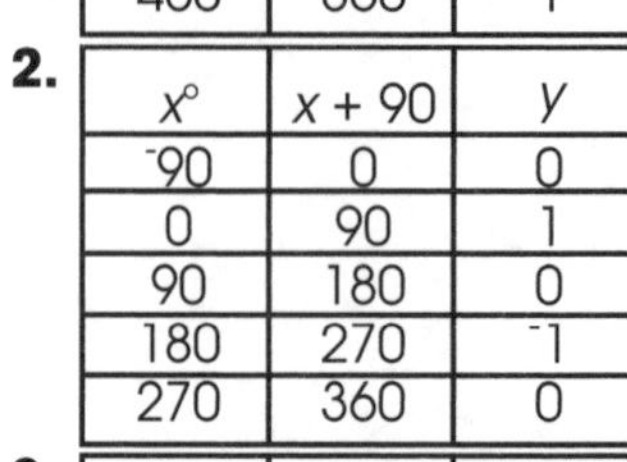

$x°$	$x + 90$	y
-90	0	0
0	90	1
90	180	0
180	270	-1
270	360	0

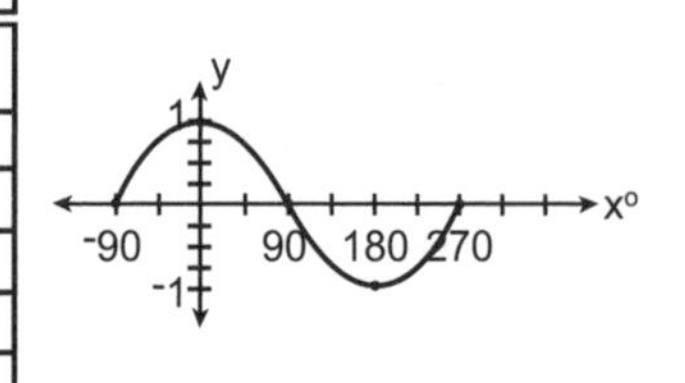

3.

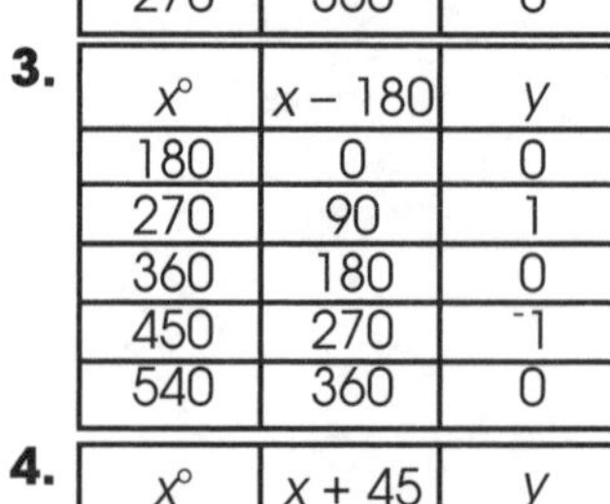

$x°$	$x - 180$	y
180	0	0
270	90	1
360	180	0
450	270	-1
540	360	0

4.

$x°$	$x + 45$	y
-45	0	1
45	90	0
135	180	-1
225	270	0
315	360	1

5.

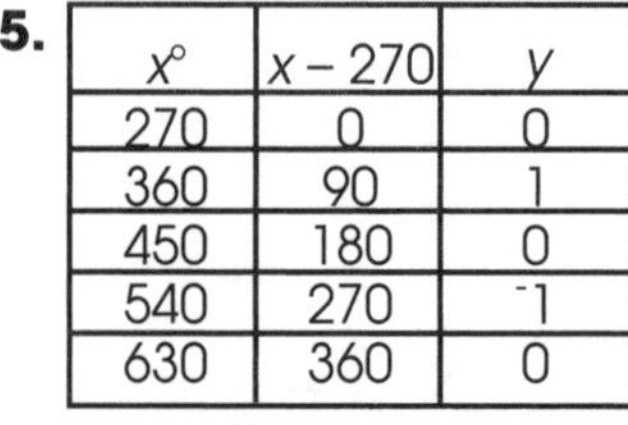

$x°$	$x - 270$	y
270	0	0
360	90	1
450	180	0
540	270	-1
630	360	0

6.

$x°$	$x + 30$	y
-30	0	1
60	90	0
150	180	-1
240	270	0
330	360	1

Answer Key

Page 78

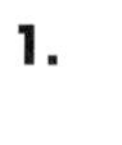

1.

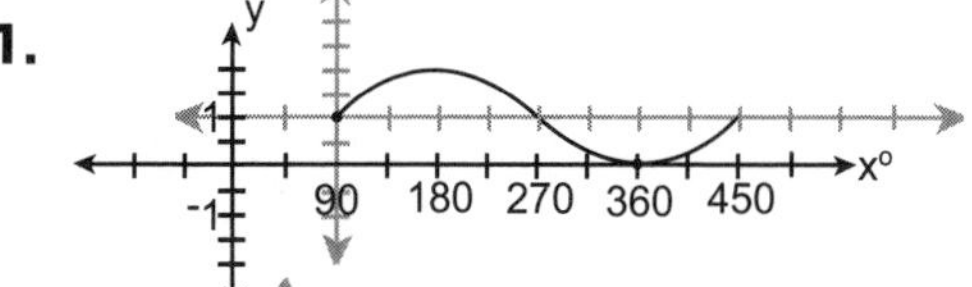

2.

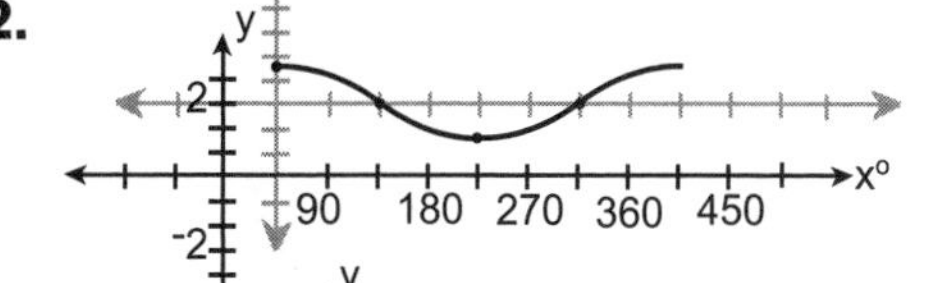

3.

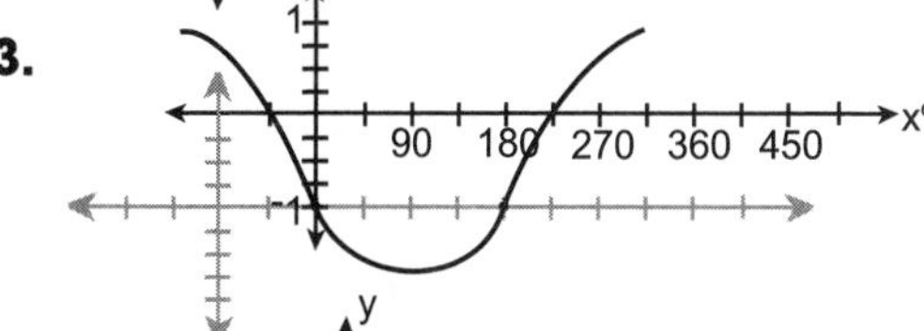

4.

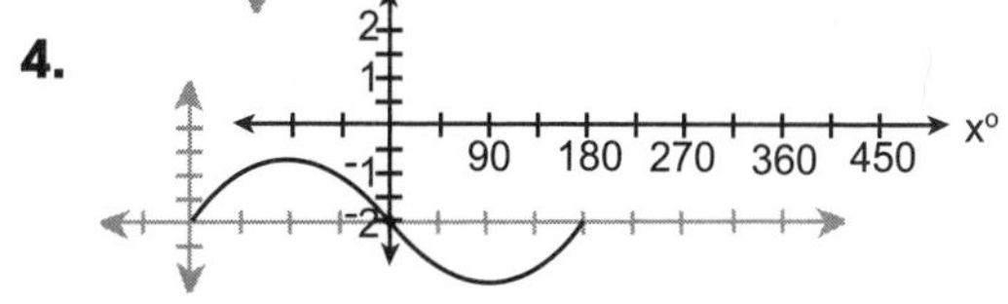

Page 80

1.

$x°$	$3x$	y
0	0	-1
30	90	0
60	180	1
90	270	0
120	360	-1

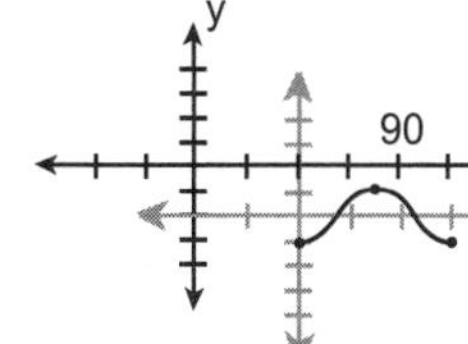

2.

$x°$	$2x$	y
0	0	1/2
45	90	0
90	180	-1/2
135	270	0
180	360	1/2

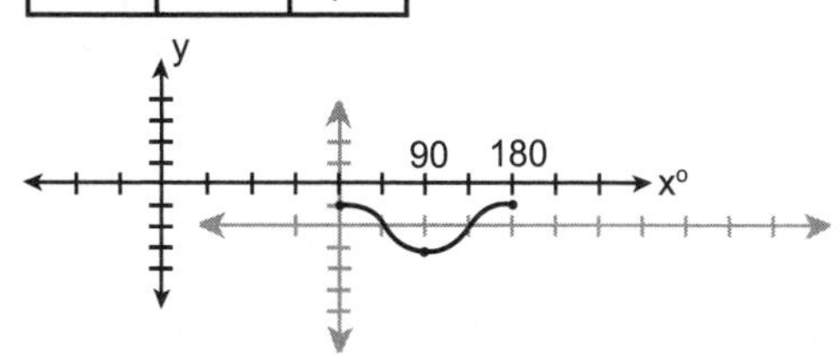

3.

$x°$	x	y
0	0	0
90	90	3
180	180	0
270	270	-3
360	360	0

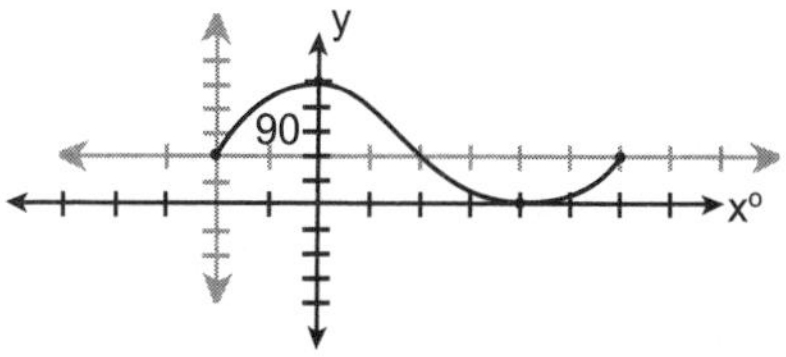

4.

$x°$	$1/2x$	y
0	0	0
180	90	-2
360	180	0
540	270	2
720	360	0

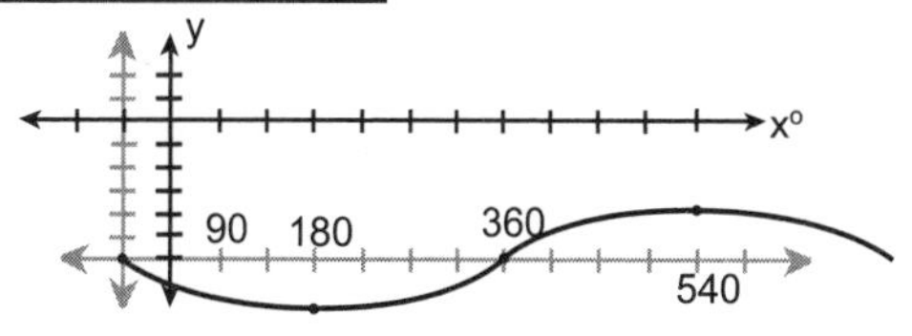

5.

$x°$	$4x$	y
0	0	3
22.5	90	0
45	180	-3
67.5	270	0
90	360	3

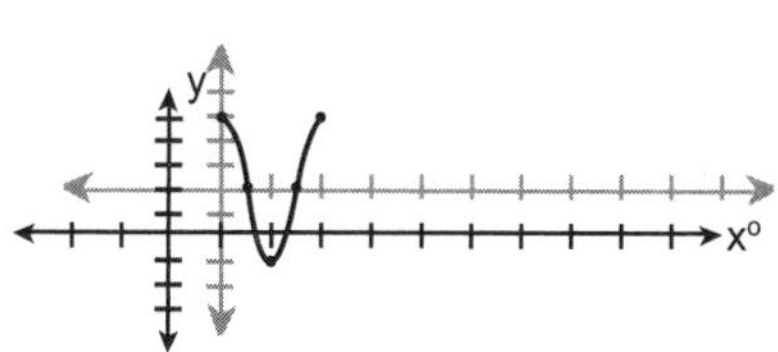

Answer Key

Page 81

1. $C = 21°$	$A = 110°$	$a = 119.5$
2. $A = 22°$	$a = 6.5$	$c = 17.1$
3. $A = 61°$	$B = 54°$	$a = 8.7$
4. $A = 22°$	$B = 48°$	$b = 7.9$
5. $C = 30°$	$c = 15$	$b = 26$ or $15\sqrt{3}$
6. $C = 39°$	$c = 28.4$	$b = 44.3$
7. $A = 60°$	$a = 12$	$c = 12$
8. $B = 15°$	$A = 126°$	$a = 15.4$

Page 82

1. $x = 132°$ **2.** $x = 19$ **3.** $x = 73°$
4. $x = 12$ **5.** $x = 26$ **6.** $x = 22°$

Page 83

1. $\dfrac{\sin 100}{10} = \dfrac{\sin 45}{x}$ $x = \dfrac{10 \sin 45}{\sin 100} = 7.2$ miles

2. $x = 200$ feet

3. $c = \cos^{-1}\left(\dfrac{153^2 - 201^2 - 175°}{^-2(201)(175)}\right) = 47°$

4. $x = \sqrt{2^2 + 4^2 - 2(2)(4)\cos 45°} = 2.8$ miles

5. $\dfrac{\sin 50°}{20} = \dfrac{\sin 45°}{x}$ $x = \dfrac{20 \sin 45°}{\sin 50°} = 18.5$ miles

Page 84

1.

x	y
5	$^-1$
5	$^-3$

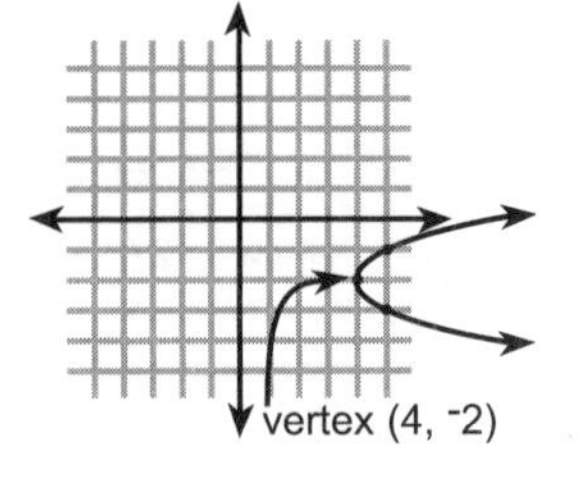

2.

x	y
$^-5$	4
$^-5$	0

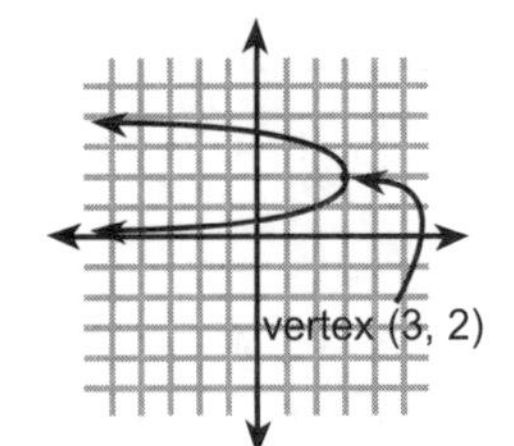

3.

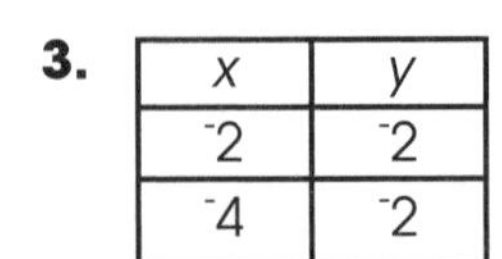

x	y
$^-2$	$^-2$
$^-4$	$^-2$

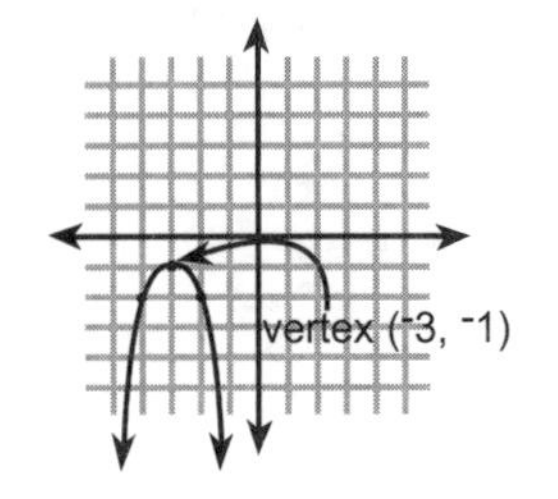

4.

x	y
$^-2$	3
$^-8$	3

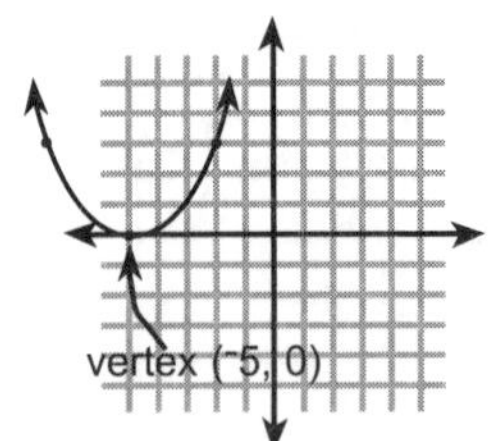

5.

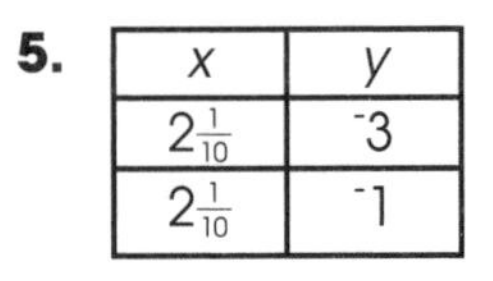

x	y
$2\frac{1}{10}$	$^-3$
$2\frac{1}{10}$	$^-1$

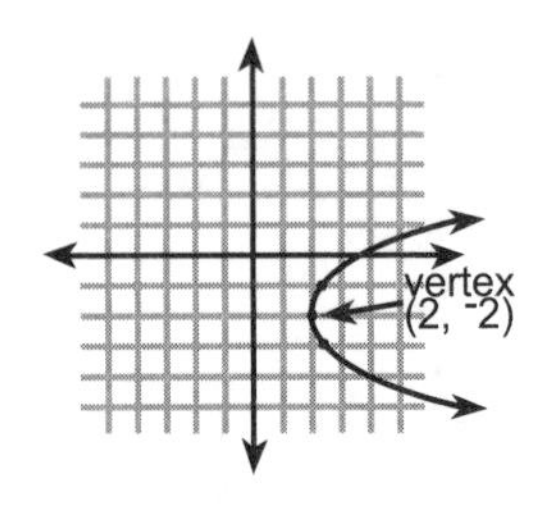

Page 85

1. $y = 2(x - 1)^2 + 6$
$v = (1, 6)$

2. $y = {}^-3(x + 2)^2 - 1$
$v = ({}^-2, {}^-1)$

3. $y = \dfrac{1}{3}(x - 3)^2$
$v = (3, 0)$

4. $y = \dfrac{1}{5}(x - 1)^2 + 2$
$v = (1, 2)$

5. $x = (y + 5)^2 - 31$
$v = ({}^-31, {}^-5)$

6. $x = (y - 5)^2 + 10$
$v = (10, 5)$

7. $x = 5(y + 4)^2 - 3$
$v = ({}^-3, {}^-4)$

8. $x = \dfrac{1}{2}\left(y - \dfrac{3}{2}\right) - \dfrac{11}{8}$
$v = \left(\dfrac{^-11}{8}, \dfrac{3}{2}\right)$

Answer Key

Page 86

1. center (4, -10)
   radius = 12
2. center (0, 7)
   radius = 7
3. center (0, 0)
   radius = 1
4. center (-3, -11)
   radius = $\sqrt{15}$
5. center (15, 0)
   radius = $\sqrt{10}$

1. $x^2 + y^2 = 64$
2. $(x + 2)^2 + (y - 3)^2 = 4$
3. $(x + 7)^2 + (y + 18)^2 = 196$
4. $(x - 12)^2 + (y - 9)^2 = 1$
5. $(x - 10)^2 + y^2 = 484$

6. center (0, 12)
   radius = $\sqrt{20}$

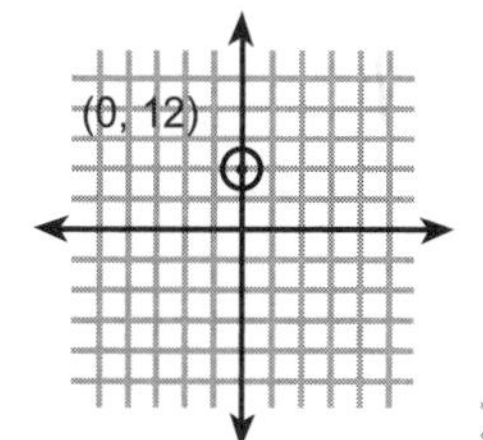

7. center (-6, -9)
   radius = $\sqrt{15}$

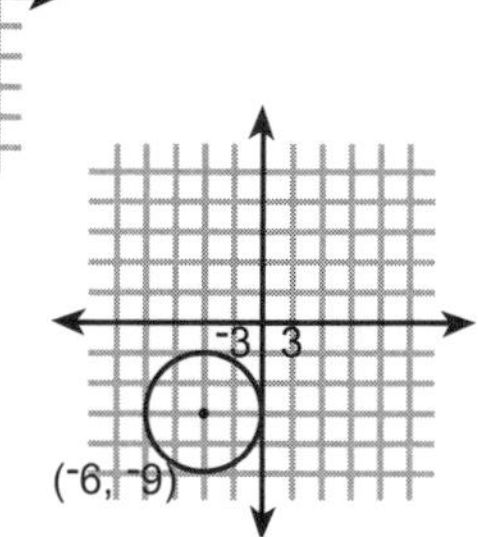

Page 87

1. center (0, 3)
   radius = 4

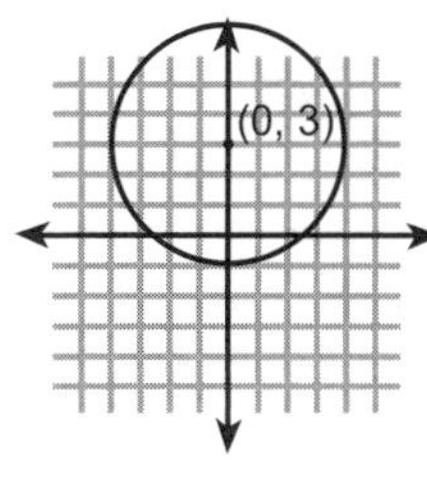

2. center (0, 0)
   radius = 8

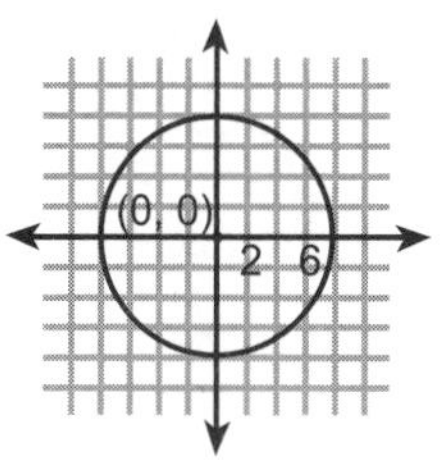

3. center (1, -1)
   radius = 1

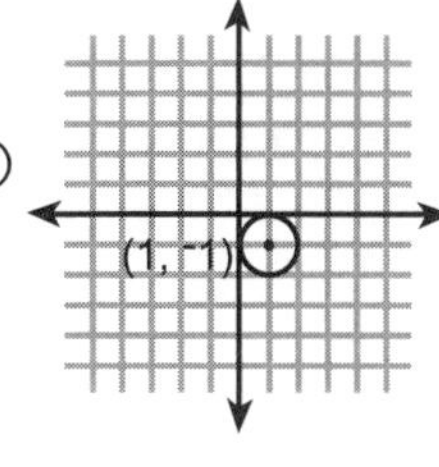

4. center (7, 2)
   radius = 5

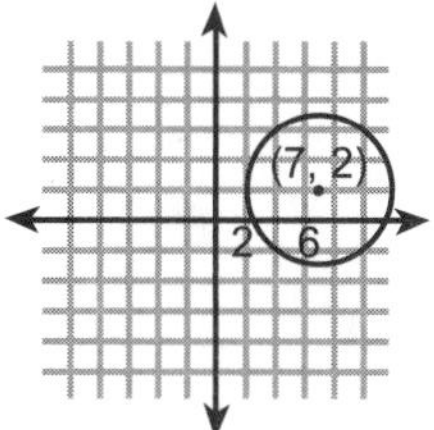

5. center (-4, 0)
   radius = 3

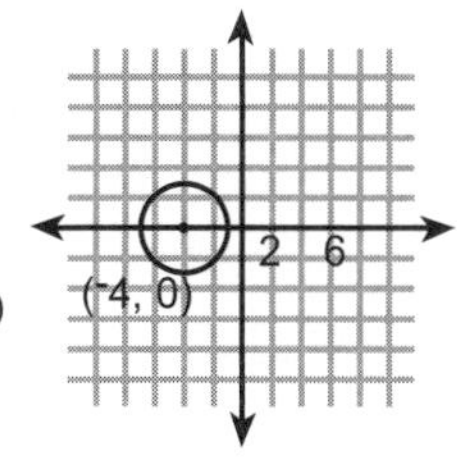

Page 89

1. $c = \sqrt{25 - 4} = \sqrt{21}$
   $c \approx \pm 4.58$

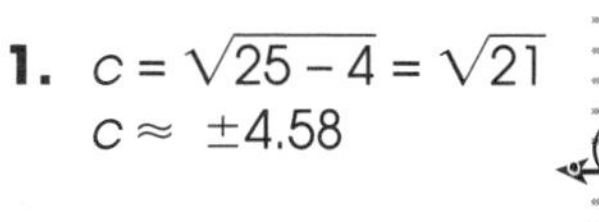

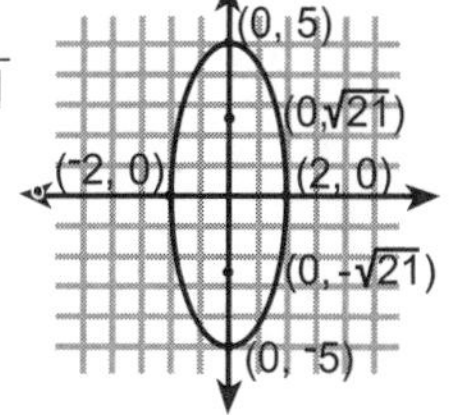

2. $c = \sqrt{64 - 1} = \sqrt{63}$
   $c \approx \pm 7.93$

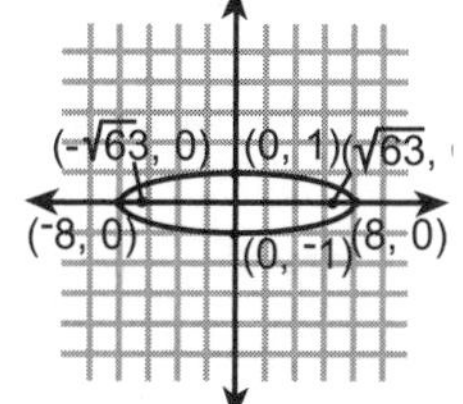

3. $c = \sqrt{64} = 8$

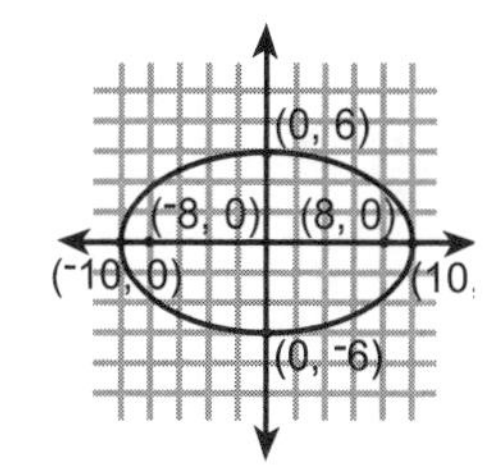

4. $c = \sqrt{95}$
$c \approx \pm 9.75$

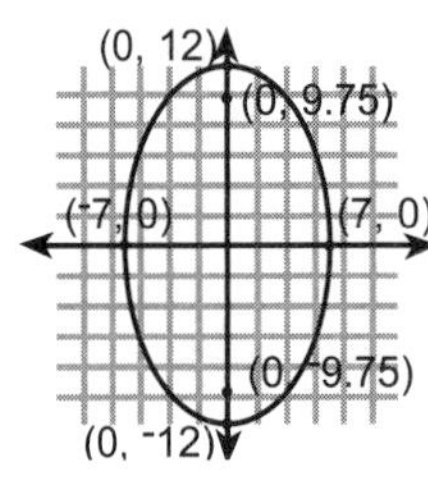

5. $c = \pm\sqrt{24}$
foci: (0, 24)
$(0, -\sqrt{24})$

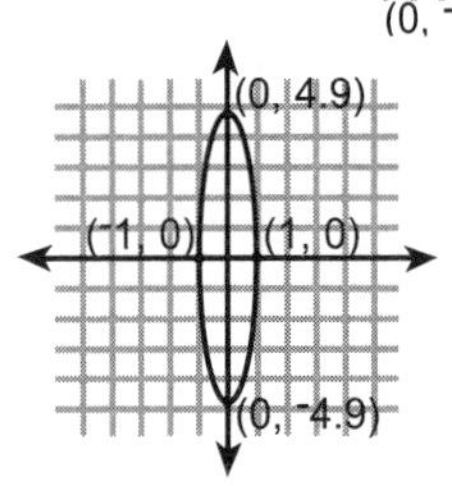

6. $c = \pm\sqrt{105}$
foci: $(0, \sqrt{105})$
$(0, -\sqrt{105})$

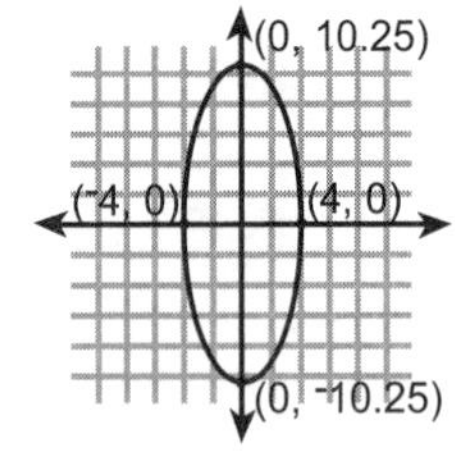

Page 90

1. center (3, 5)

2. center (⁻8, 6)

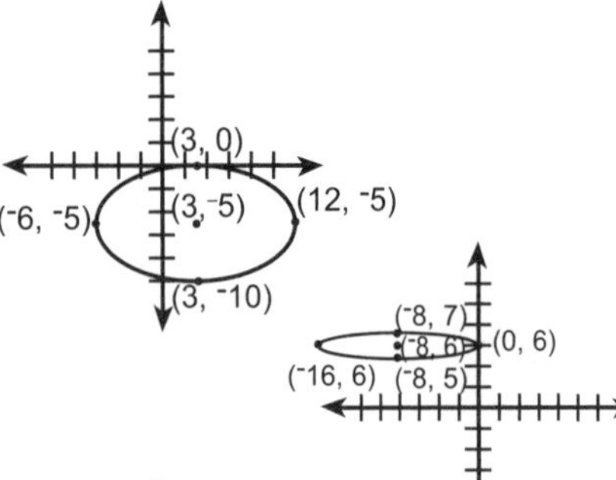

3. center (0, 0)

4. center (⁻2, ⁻3)

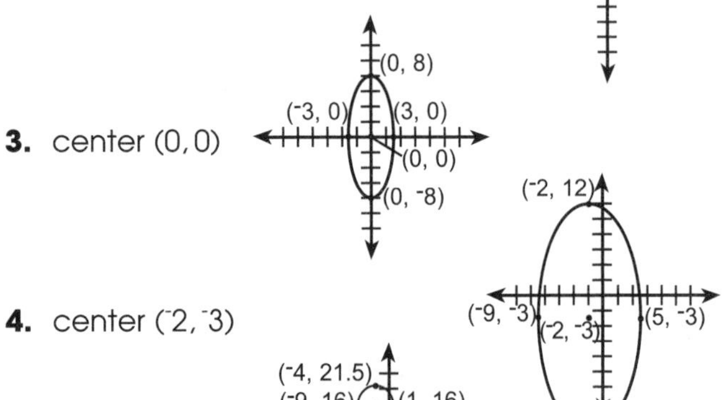

5. center (⁻4, 16)

6. center (5, 11)

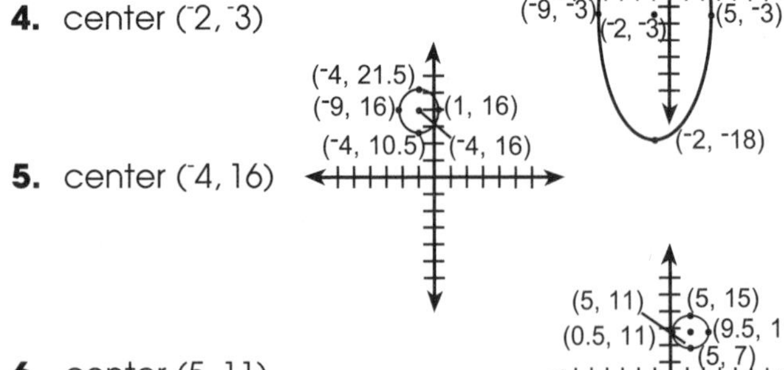

Page 92

1. foci $(\sqrt{74}, 0)$
$(-\sqrt{74}, 0)$

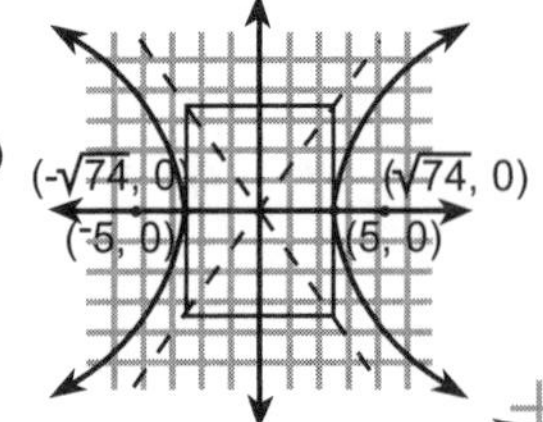

2. foci $(0, \sqrt{97})$
$(0, -\sqrt{97})$

3. foci $(0, \sqrt{136})$
$(0, -\sqrt{136})$

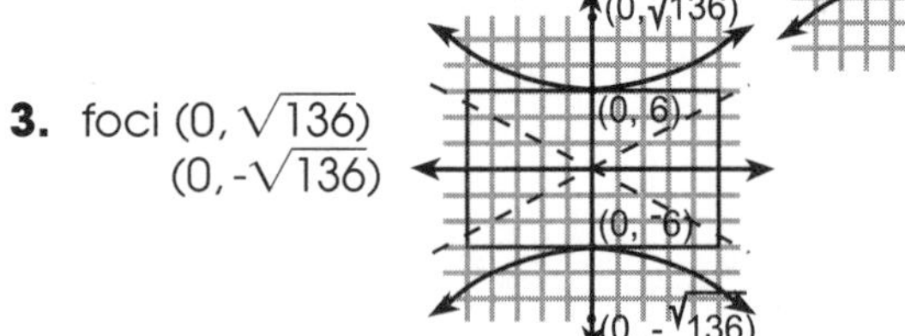

4. foci $(\sqrt{193}, 0)$
$(-\sqrt{193}, 0)$

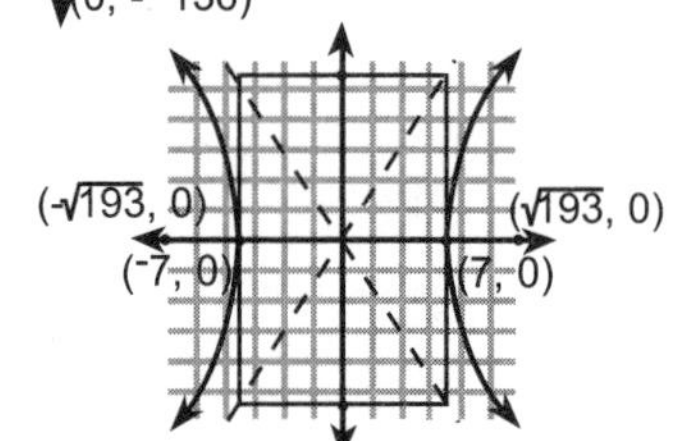

5. foci $(\sqrt{26}, 0)$
$(-\sqrt{26}, 0)$

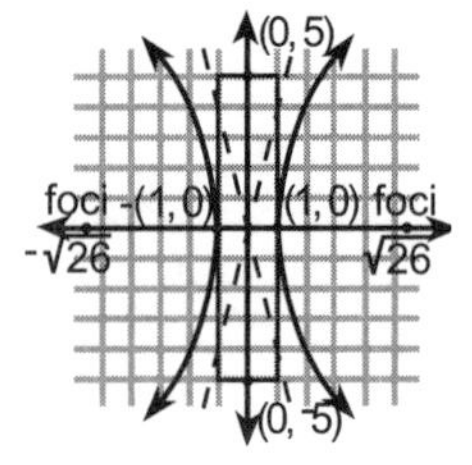

6. foci $(0, \sqrt{137})$
$(0, -\sqrt{137})$

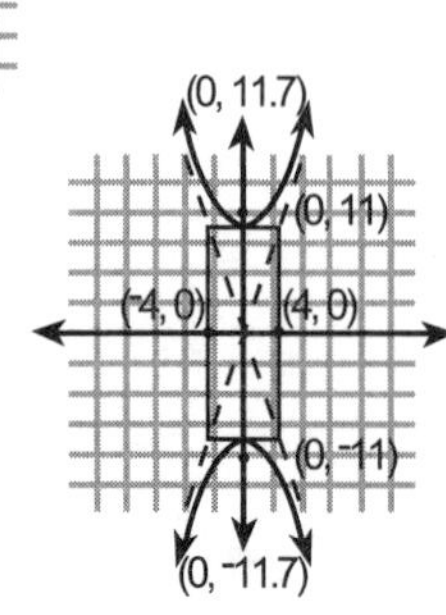

Answer Key

Page 93

1. center (-3, -5)

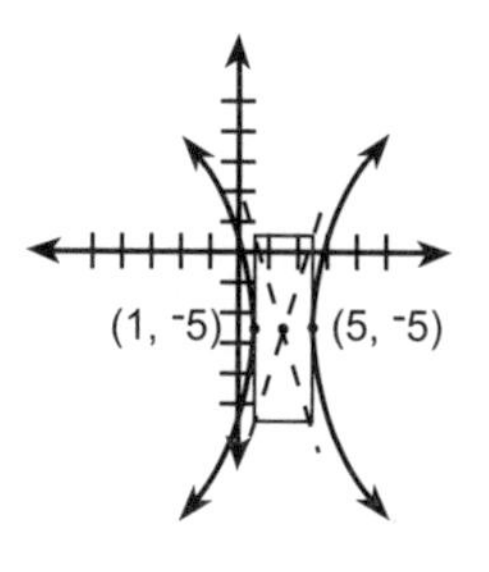

2. center (-8, -6)

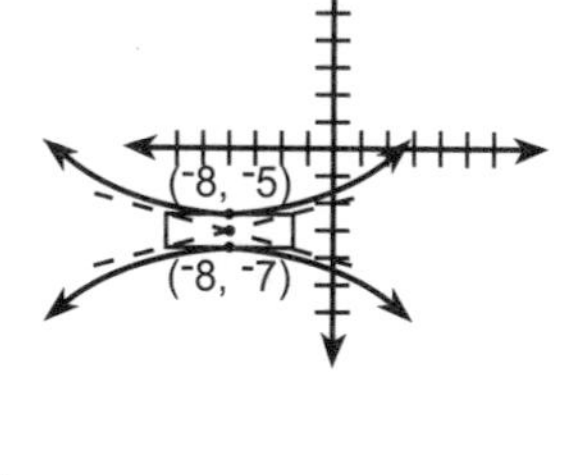

3. center (0, 0)

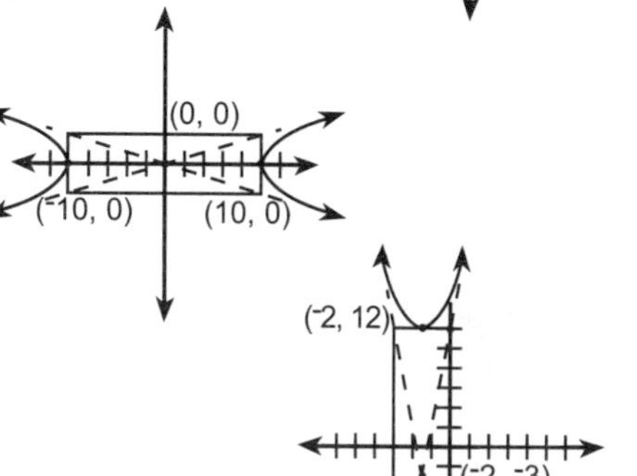

4. center (-2, -3)

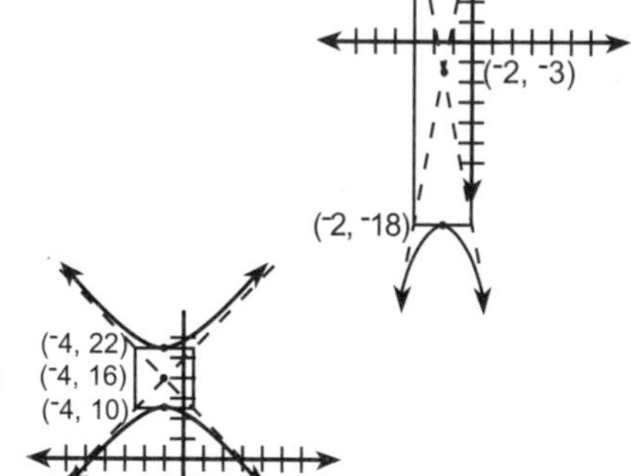

5. center (-4, 16)
   vert: (-4, 22)
   (-4, 10)

6. center (5, 11)
   vert: (9, 11)
   (1, 11)

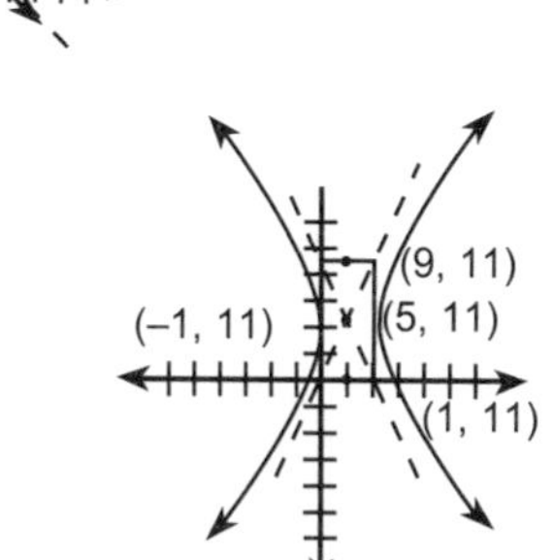

Page 94

1. $(x-2)^2 + (y+3)^2 = 9$

This is the equation of a circle with a center at (2, -3).

2. $\frac{(x+3)^2}{4} + \frac{(y+1)^2}{9} = 1$

This is an ellipse with a center at (-3, -1).

3. $\frac{(x-6)^2}{1} + \frac{(y+4)^2}{25} = 1$

This is an ellipse with a center at (6, -4).

4. $\frac{(x+3)^2}{1} - \frac{(y-4)^2}{16} = 1$

This is the equation of a hyperbola with a center at (-3, 4).

5. $(x+4)^2 + (y+10)^2 = 4$

This is the equation of a circle with a center at (-4, -10).

Page 95

1. center (0, 2)

2. center (0, 3)

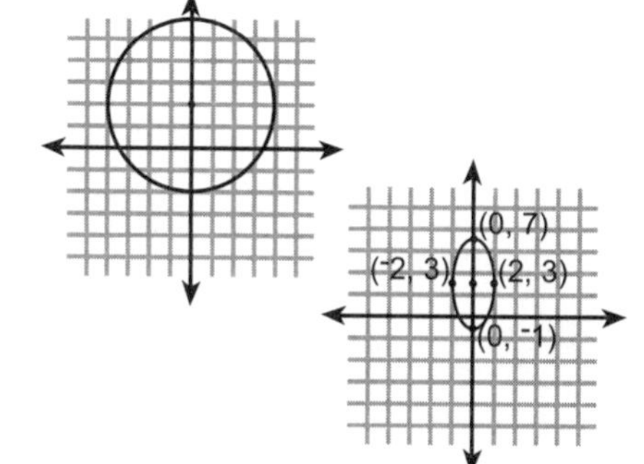

3. center (1, -5)

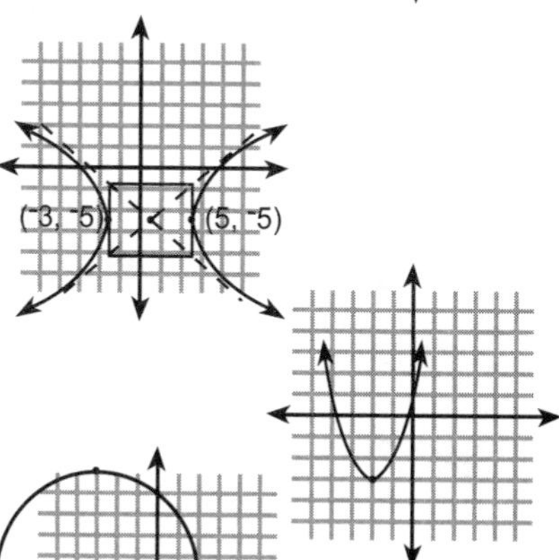

4. vertex (-2, -3)

5. center (-3, 1)

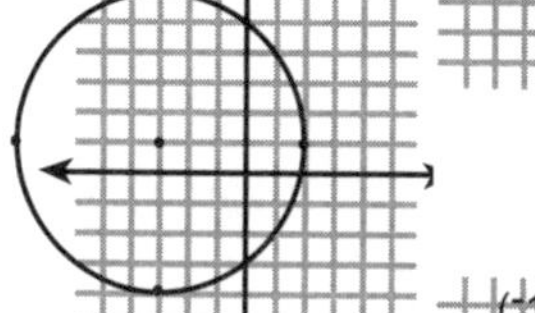

6. center (-1, 3)

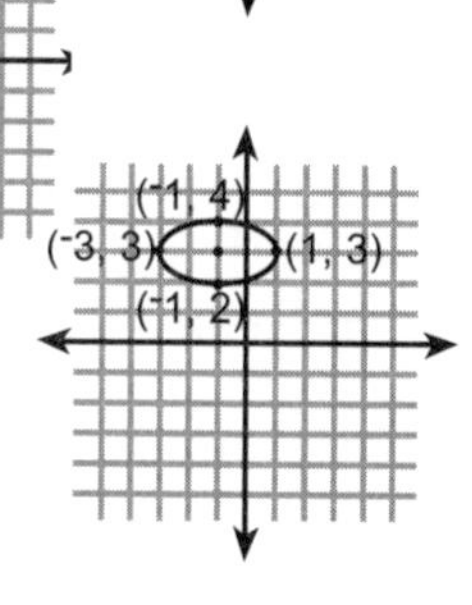

Answer Key

7. center (-2, 1)

8. vertex (-4, 4)

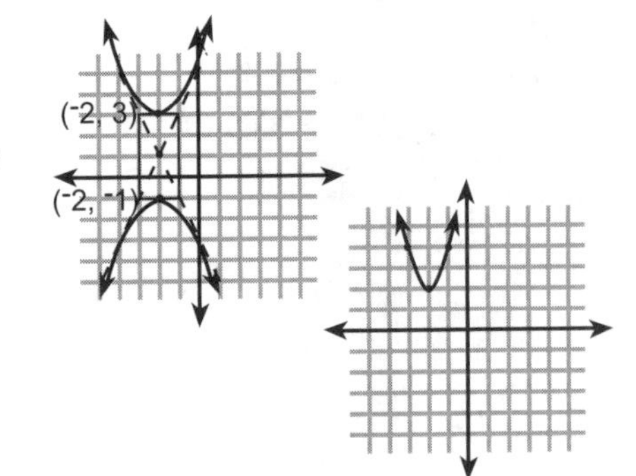

Page 96

1. 11 meters at 49° west of north
2. 7 meters at 25° south of west
3. 12 meters at 65° north of east
4. 25 meters at 70° east of south
5. 14 meters at 55° west of south
6. 23 meters at 37° east of north

Page 97

1. 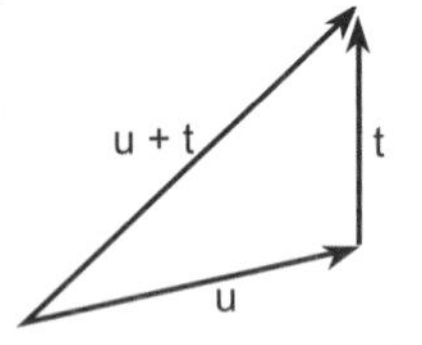

2. 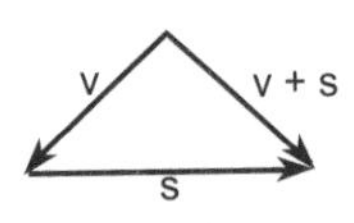

3. 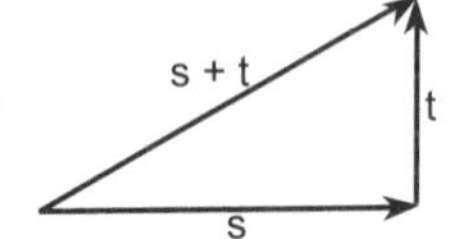

4. 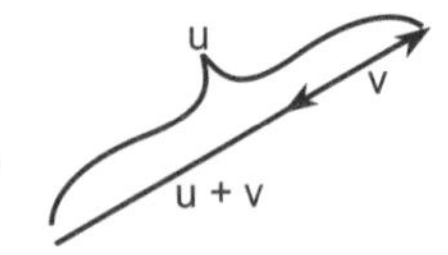

5. 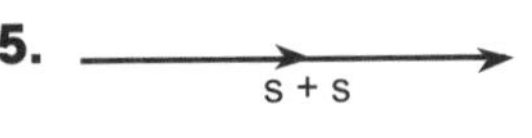

6. 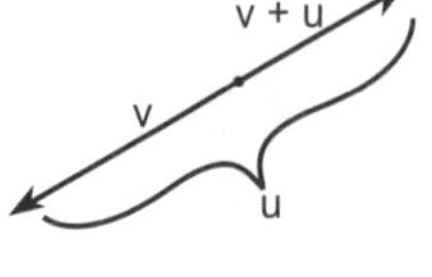

7. 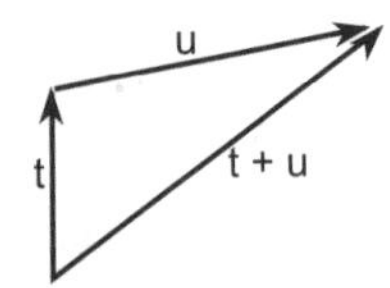

8. 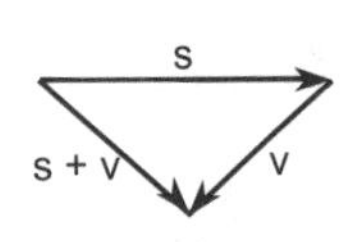

Yes, either way you add, the resulting vector is the same.

Page 98

1. 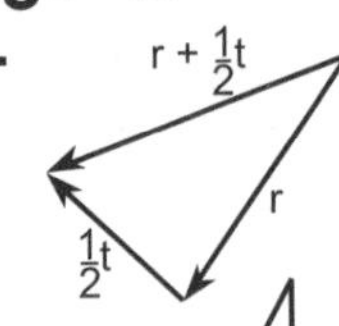

2. 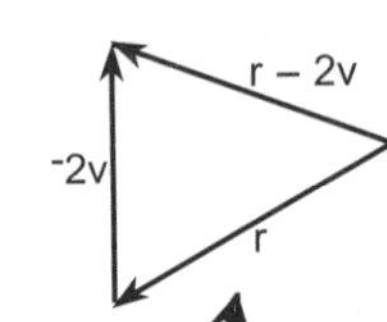

3. 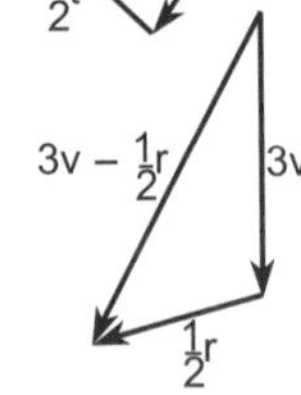

4. 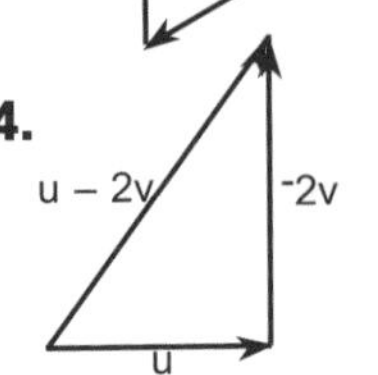

5. 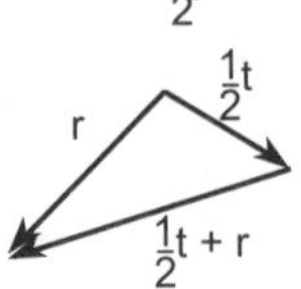

6. 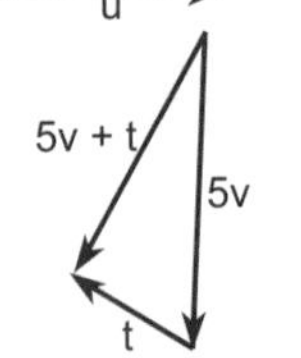

7. 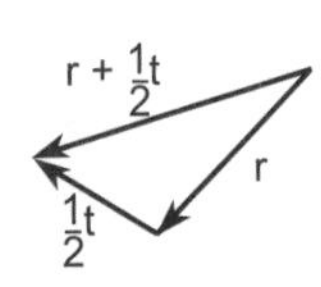

8. 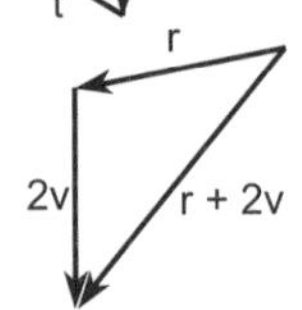

No, when you subtract the opposite vectors, you are changing the direction of the resulting vector.

Answer Key

Page 99

1. $x = {}^{-}8.3\text{m} \quad y = 7.2\text{m}$ **2.** $y = {}^{-}6.3\text{m} \quad x = {}^{-}3.0\text{m}$
3. $x = 5.1\text{m} \quad y = 10.9\text{m}$ **4.** $x = 23.5\text{m} \quad y = {}^{-}8.6\text{m}$
5. $x = {}^{-}11.5\text{m} \quad y = {}^{-}8.0\text{m}$ **6.** $x = 13.8\text{m} \quad y = 18.4\text{m}$

Page 100

1. 17.8 meters at 52° north of east
2. 16.6 meters at 65° south of west
3. 25.9 meters at 62° south of east
4. 40.6 meters at 52° north of west
5. 10.2 meters at 11° south of east
6. 4.1 meters at 76° south of west

Page 101

1. 30.4 mph at 9°
2. He must turn 8.5° east of south
3. 18.9 miles 58° north of west
4. 70 meters north; 70 meters west
5. 201 miles south

Page 102

1. 18.3m at 32° north of east
2. 25.6m at 17° south of east
3. 101.9m at 22° south of west
4. 10.8m at 46° north of west
5. 83.9m at 3° north of east

Page 103

1. 35 mph at 51° north of west
2. 9.4 mph at 67° north of west
3. 13.3 miles 78° north of west
4. 444.4 mph at 43° north of east
5. 3.2 mph at 2° south of east

Notes